# Drainage of Highway and Airfield Pavements

# Drainage of Highway and Airfield Pavements

HARRY R. CEDERGREN

A Wiley-Interscience Publication

JOHN WILEY & SONS New York • London • Sydney • Toronto

***Library of Congress Cataloging in Publication Data:***

Cedergren, Harry R.
Drainage of highway and airfield pavements.

"A Wiley-Interscience publication."
Includes bibliographies.
1. Road drainage. 2. Airport drainage.
3. Pavements—Design and construction. I. Title.

TE215.C36 625.7'34 74-13400
ISBN 0-471-14181-X

Printed in the United States of America

10 9 8 7 6 5 4 3 2 1

To the memory of JOHN McADAM and
HENRI DARCY, whose historic efforts
provide the foundations on which
this book was developed

# Preface

For centuries road builders have known that water accelerates the deterioration and destruction of pavements, and they have expressed the belief that good drainage is one of the basic needs of pavements. In spite of this knowledge, however, well-drained pavements are exceedingly scarce. Inasmuch as pavement designs are developed from tests on saturated specimens of subgrades and bases, it has been assumed that any and all effects of water will be taken into account automatically. Nevertheless, there is mounting evidence that this "undrainage" philosophy and its procedures do not allow fully for the damaging effects of dynamic wheel loads on structural sections containing excess water, for many important pavements have been breaking up faster than can be attributed solely to increasing volumes of traffic.

For the past several decades the tendency has been to consider that environmental factors such as rainfall do not need to be allowed for in pavement design. Nevertheless, it is well known that excess water is almost always in evidence during the times that deterioration and failure are occurring most rapidly.

The untimely repair and replacement of many miles of prematurely deteriorated pavements is placing increasing demands on diminishing supplies of natural resources, which emphasizes the need to return to the good drainage philosophy expressed by John L. McAdam more than 150 years ago.

In the period from 1970 to 1973, I made field surveys and tests of interstate freeways, state highways, and airfield runways, taxiways, and aprons in widespread regions of the United States to study and evaluate the possibility that water (and surface water in particular) may be responsible for the accelerating rates of damage to pavements. These surveys and tests, together with my personal observations over the past 35 years, have confirmed my belief that slow drainage of water out of pavements is all too often a major factor in their premature deterioration, and that pavements designed as rapidly draining systems should be able to carry their design loads almost indefinitely with only normal, routine maintenance, whereas their undrained counterparts would require large amounts of costly repairs or replacement long before reaching their design age.

This book reviews the basic fundamentals of seepage and infiltration into structural sections and flow of water out of drainage systems. It presents methods for evaluating the damage caused by heavy wheel impacts on pavements containing excess water, and the benefits of rapid drainage. It gives basic criteria needed for designing drainage systems, with several examples of appropriate types. Also, it contains numerical examples of the analysis and design of these systems.

In developing this book I have drawn extensively from my own experiences, but I have also made use of research by outstanding thinkers and workers in the field of pavement design and drainage. It would be difficult to thank all of those who have contributed to the advancement of this important field, but wherever published information has been used, appropriate references have been made. Special appreciation is given, however, to the Federal Highway Administration (FHWA), Office of Research, Washington, D.C., and to the U.S. Army Corps of Engineers, Washington, D.C., Champaign, Illinois, Vicksburg, Mississippi, and Hanover, New Hampshire.

This book has been developed in the hope that by assembling this information in one volume more attention will be focused on the one factor above all others that damages pavements and that it will help to dispel the concept that excess water is something that must be lived with.

No attempt has been made to separate airfield and highway pavements into two or more different categories of structural types, since the major differences in pavements for various uses are in their dimensions and proportions, but the problems with water are essentially alike.

I sincerely believe that the application of principles brought out in this book will conserve natural resources and offer substantial long-term savings to those who pay for pavements, while providing facilities that will be more serviceable, last longer, and give more comfort and safety to the users.

*Harry R. Cedergren*

*Sacramento, California*
*August 1974*

# Contents

# Notation

$A$ = area, dimension
$a$ = area, dimension
$B$ = width
$b$ = dimension
$C$ = coefficient
$c$ = dimension
$D$ = dimension
$D_2$ = 2 percent size (2 percent is finer)
$D_{15}$ = 15 percent size (15 percent is finer)
$D_{50}$ = 50 percent size (50 percent is finer)
$D_{85}$ = 85 percent size (85 percent is finer)
$d$ = dimension
$g$ = longitudinal grade of a pavement
$H$ = dimension
$h$ = hydrostatic head, dimension
$I$ = inflow, infiltration, design infiltration rate
$I_g$ = equivalent infiltration (groundwater inflow)
$i$ = hydraulic gradient
$k$ = coefficient of permeability
$k_b$ = permeability of porous base
$k_g$ = "global permeability" (effective permeability of a pavement area, including cracks, joints, etc.)
$L$ = distance, length
$n$ = porosity
$n_d$ = number of equipotential drops in a flow net
$n_e$ = effective porosity
$n_f$ = number of flow channels in a flow net
$O$ = seepage outflow, water loss
$Q$ = seepage quantity
$q$ = seepage quantity
$S$ = distance
$S$ = serviceability, severity factor (percent)
$s$ = distance, slope

$T$ = time; effective pavement life, years
$T$ = dimension
$t$ = time, age, years
$t_{50}$ = time for 50 percent drainage
$t$ = dimension, clear thickness of open-graded drainage layer
$t_b$ = thickness of open-graded drainage layer
$v_d$ = discharge velocity
$v_s$ = seepage velocity
$W$ = width, length
$w$ = maximum drainage path
$x$, $z$ = cartesian coordinates

# Drainage of Highway and Airfield Pavements

CHAPTER ONE

# Introduction

## 1.1 SIZE AND SCOPE OF THE PROBLEM

Within the continental United States there is a 43,000-mile system of interstate highways (nearly completed in 1974), another 500,000 miles of surfaced state highways, around 500,000 miles of city streets, about 1.8 million miles of improved county and township roads, and large expanses of airfield pavements, parking lots, and miscellaneous pavements. According to a Highway Research Board report, the mileage of paved roads in the United States totaled 2,827,000 in 1967. Throughout all developed countries the total mileage is probably in the order of 8 or 10 million. These vast networks of pavements represent a large investment by taxpayers, probably adding up to a total of $400 to $500 billion. Figure 1.1 shows some of the kinds of pavements that have been constructed in the United States in the past 10 to 20 years. Many of them add to the beauty of the land on which they are built. Figure 1.1*a* is a modern, multilane freeway in Detroit Michigan; Fig. 1.1*b* is a view along the centerline of a taxiway of a major airport in a western state;[1] Fig. 1.1*c* is a view along a newly built state highway in an eastern state.

All of these pavement systems are exposed to the weathering actions of nature, which together with the pounding of traffic often combine in such ways as to lower their quality and serviceability. Routine maintenance of the world's pavements, including seal coats, joint sealing, slurry seals, armor coats, and so forth, is costing tax-supported agencies an estimated $10 million a day or around $3 billion a year. Digging out and replacing unserviceable portions of pavements of various kinds can cost another $4 million a day or about $1.2 billion a year more.

These amounts of capital investment and maintenance and repair obviously represent a great deal of interest and concern on the part of those responsible for the construction and maintenance of pavements

**FIG. 1.1** Views of modern pavements. (*a*) New, multilane freeway in Detroit, Michigan, October 1970. (*b*) Looking along centerline of airport taxiway; West coast area; September 1972. (Courtesy CERL.[1]) (*c*) New state highway; eastern state; September 1970.

[1] Wherever this notation appears in this book, it pertains to photos or other illustrative matter used in the report, "Methodology and Effectiveness of Drainage Systems for Airfield Pavements," prepared by the author under contract with the Construction Engineering Research Laboratory (CERL), U.S. Army Corps of Engineers, Champaign, Illinois, and scheduled for publication by CERL in 1974.

(c)

everywhere. And, just as obviously, any method of design or repair that can cut costs or extend the length of useful, trouble-free life of these facilities should be of considerable concern to those designing and building them as well as to those using and paying for them.

When engineers are comparing alternate pavement designs, it is important that they consider not only the original costs and routine maintenance and upkeep, but all significant strengthening costs and replacement costs over the expected useful life of the system. Here is where there is a weakness in the standard procedures, because in most cases only the original cost and "normal" maintenance are used in estimating the relative costs of alternative designs. Since the inclusion of effective subsurface drainage features in new pavements may add slightly to the original costs, they are almost never included (except for groundwater). Yet, their omission may be cutting life and increasing repair and replacement costs far more than the added cost of good drainage. Estimates presented in Chapter 8 indicate that poorly drained pavements may in many cases be more than twice as costly as well-drained pavements when compared on a long-term basis.

Most of the pavements built in the past several decades are stratified bodies composed of two or more similar or dissimilar layers, such as portland cement concrete (PCC) on stabilized or unstabilized bases, or asphalt concrete (AC) on stabilized or unstabilized bases. These systems, which generally are not designed for rapid elimination of water, have the mechanisms of self-destruction built in. When free water gets into the boundaries between structural layers, the multi-layered systems act very much as diaphragm pumps under the pounding of heavy wheel loads. For example, each heavy-load impact causes water to move about at the interface between a wearing course and its base, eroding material and ejecting it through cracks and joints, producing channels and cavities that undermine the pavement and the result is faulting and other damage that eventually leads to total failure of the pavement.

Normal design methods that develop pavement cross sections—on the assumption that the controlling factors are stress and strain, deformation, volume change, and fatigue under repeated loadings—largely ignore the dynamic effects of excess water, which can completely overshadow the factors that *are* considered. Thus erosion between a primary pavement layer and its base can occur regardless of the thickness or rigidity of the base, if conditions conducive to erosion exist at their interface. Erosion, damaging actions of pore pressures, and so on *cannot occur in the absence of free water*; hence there is reason to believe that *well-designed, well-drained pavements should carry their design loads almost indefinitely with only normal, routine maintenance, whereas their undrained counterparts may require large amounts of costly repairs long before their design age* is reached.

Pavement construction contracts generally provide for subdrains to take care of high groundwater, spring inflows, and the like that are either known about in advance of construction or discovered during the conduct of the work. There is, however, a widespread philosophy that infiltration from rainfall and other surface sources does not need to be (or cannot be) drained from structural sections; thus drainage systems are almost never provided for this purpose. Although groundwater inflows certainly need to be controlled, it is unlikely that more than about 5 percent of the total pavements in the United States need this kind of protection. *On the other hand, at least 90 to 95 percent of the pavements in this country are in areas where rainfall rates are greater than subgrade drainage capabilities, and very few have drainage systems that will handle this problem.*

Why does this anomalous condition exist? Probably in part, because severe groundwater conditions usually show up during construction or are revealed by advance drilling and groundwater observations, and these sources often create wet, boggy conditions that must be controlled to make the construction possible. In contrast, surface water infiltration occurs after a pavement is completed and put in service. It exerts influences on pavement condition and serviceability that are *hidden beneath the surface* and do not necessarily show up instantly. Although surface water infiltration may eventually cut the life of a given pavement to half or less of its potential were it well drained, this is a gradual process, and during the first few years its effects may go unnoticed. The only outward evidence for several years may be the dry residues showing up as stains near cracks and joints and across shoulders, and the faulting at joints and cracks. Figure 1.2*a*, for example, shows water bleeding out at the shoulder joint of a 2-year-old interstate highway on a 50-ft-high fill several days after a rain. The traveled way is paved with 8 in. of PCC on 3 to 4 in. of cement-treated granular base over an aggregate subbase. Whenever water can be seen flowing out of joints and across shoulders, as in this photograph, there must be sufficient head or pressure *under the adjacent pavement* to lift the water out (with or without the help of the impacts of heavy wheel loads). In many cases, it is forced out by a *combination* of artesian head and the pressures of traffic.

In order for the condition in Fig. 1.2*a* to exist, any cavities, void spaces, and so forth between the pavement and its base must be full of water, which puts the structural section in a virtual "bathtub." Because the compacted subgrades of most of the pavements in the United States have coefficients of permeability lower than rainfall rates that occur every year, the majority of these pavements have to carry traffic while full of water for significant amounts of time each year. And, the existence of this exposure to excess water is virtually independent of whether a given road or other pavement is in a cut or high on a fill—as in Fig. 1.2*a*—since it can be caused by precipitation.

Just how important is this so-called "bathtub" condition? Several experimental road tests which have been made since 1950 have documented records that indicate that during the periods when the pavements contained large amounts of free water the rates of deterioration in the various tests ranged from 10 or 20 up to hundreds and thousands of times greater than during the times when they contained little or no free water (see Table 2.1). At the site of the photograph in Fig. 1.2*a*, another picture was taken from a slightly different point 10 years later. This picture, Fig. 1.2*b*, shows that extensive repairs and damage have already taken place, at an age of 12 years, although the pavement should have given good service for at least 18 to 20 years without major repairs.

Figure 1.3*a* shows an asphalt concrete surfaced state highway in an eastern state that has been severely damaged after fewer than 10 years

**FIG. 1.2** Interstate highway on 50-ft high fill; poorly drained roadbed stays in "bathtub" condition from surface infiltration long after each rain. (*a*) Two years after construction; telltale signs of internal flooding; April 1960. (*b*) Twelve years after construction; excess water and traffic are taking their toll; December 1970

of service. This section of road is on a shallow fill, with no nearby sidehill cuts of any depth; hence all the water that gets into its structural section is from rainfall or other forms of precipitation. Another poorly drained pavement that has been in service less than 10 years is shown in Fig. 1.3*b*. Here, the upper layers of the asphalt concrete are breaking out, while free water is filling the cavities in the pavement. This kind of "break-out" damage occurs almost entirely when free water is present at the interfaces between successive lifts of asphalt concrete pavement material or between pavement layers and bases. The pore pressures that build up in the water under traffic impacts are largely responsible for the lifting of upper layers off lower layers.

These are but a few examples of pavements that have been damaged by surface-water infiltration. An indication of the potential magnitude of the damage to modern pavements, and the mounting cost of keeping heavily used pavements in serviceable condition on a national scale, is given in a General Accounting Office (GAO) report to Congress in July 1970 (see also Chapter 8). An article in *Civil Engineering* magazine (ASCE, 1970), says that surveys made of pavements put down prior to October 1963 indicated some 2800 miles of these interstate pavements already needed overlays at an estimated cost of $200 million. By 1970 it was reported that $175 million would still be needed to upgrade parts of the 2800 miles still in need of repair. The GAO report says that engineers did not anticipate the rapid growth in truck-traffic volume, and the resultant damage to pavements. Although increased traffic volume undoubtedly did bring on many of these problems earlier than would have occurred under less traffic, the undrained condition of many of the pavements certainly must be regarded as a major factor in their premature deterioration. If these pavements had been provided with the drainage systems advocated in this book (coarse, open-graded bases with pipe outlets), it is considered unlikely that many of them would have needed more than routine maintenance up to the present time.

In times past some of the leading road builders, such as Pierre Marie Tresaguet and John McAdam, must have instinctively realized that coarse stone and gravel are relatively good conductors of seepage and infiltered surface water. Engineers of those times who understood soil behavior placed a thin layer of dry stone screenings on subgrades to prevent the subgrade soil from working up into coarse base materials (Hewes and Oglesby, 1954). And there were those who advocated the use of coarse stone for "French drains." When coarse materials were

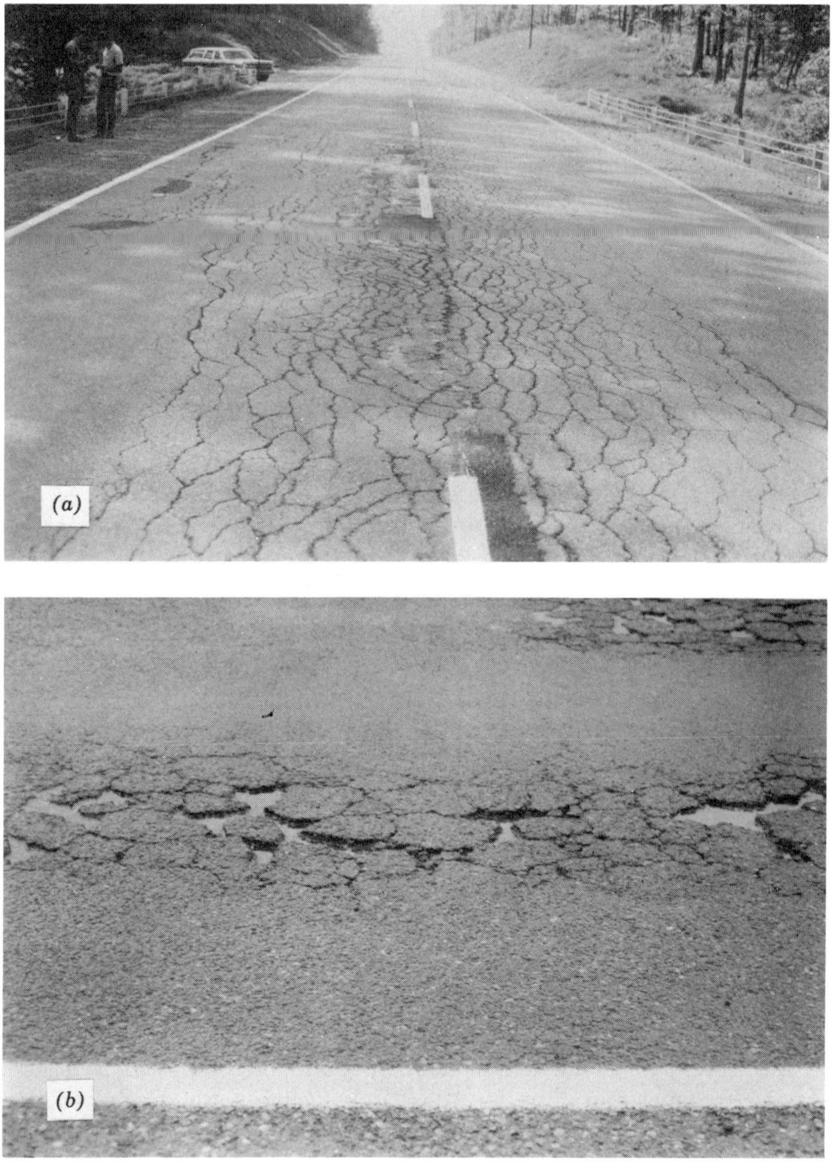

**FIG. 1.3** Asphalt concrete pavements badly damaged by traffic and surface infiltration. (*a*) Eight-year-old state highway on 10-ft fill; eastern state; September 1970. (*b*) Poorly drained AC on CTB; in Pacific Northwest; Oct., 1970; about 8 years old.

used under favorable conditions, or with the use of choker layers to keep the fine soil out, they had a good record of success. But, unfortunately, there developed a tendency to place coarse materials (both in roadbeds and in trench drains) without the proper regard for the danger of infiltration when these materials were in contact with wet, erodible sand, silt, soft clay, and so forth. As a result of this disregard for physical conditions, coarse stone and gravel bases and drains often became clogged with fine soil or mud, and the overlying roadways generally deteriorated from a lack of drainage.

As with all other human endeavors, design and construction practices go in cycles, like a pendulum swinging from one extreme to another. Thus after the bad experiences with open-graded materials, the trend went to the widespread usage of *well-graded mixtures of sand and gravel* for most highway and airport construction. Because of a fear of having drains become clogged by infiltration, many of the materials in widespread usage for base courses and drainage layers contain so many fines that they are in fact *preclogged by their own fines*, and as a consequence provide little or no beneficial drainage for the pavements they are supposed to protect.

## 1.2 STANDARD PAVEMENT DESIGNS DO NOT PROVIDE MUCH DRAINAGE

During the past several decades the primary emphasis in pavement design has been on density and stability rather than on drainability. Since pavement designs are based on the strengths of subgrades and supporting layers in a saturated condition (but without evaluating the dynamic effects of wheel impacts on free water that is often trapped in actual structural sections), it has been widely assumed that drainability is not important. Even though road builders for centuries proclaimed the virtues of drainage, and knew that free water in structural sections greatly accelerated the rate of damage, nearly all the important pavements that have been built in the past few decades are slowly draining systems containing free water for significant amounts of time. If after a given design chart or formula has been used for a number of years premature damage develops in the pavements, it has been common practice to modify the chart or formula to provide greater thicknesses of structural layers, increase the percentage of cement or other stabilizers, widen the base, or make other alterations in details to try to improve the design *but with no attempt to improve the drainability*. Many designers assume that it is neither practical nor economical, nor perhaps even necessary, to speed up the removal of water from structural sections. As a consequence compacted, stabilized bases and subbases of the kinds in widespread usage are nearly all low in permeability. Likewise, the materials used in shoulder construction usually act as dams or barriers that prevent the lateral movement of water out of structural sections.

Figure 1.4 shows cross sections through typical types of pavements that have been built within the past 20 to 30 years. The right half in Fig. 1.4*a* has its structural section confined within an impervious "choker" in its shoulder and an impervious subgrade. The left half in Fig. 1.4*a* is similar, but has a "daylighted" sand blanket which is considerably more permeable than the choker in the right side, but still too low in permeability to provide any significant amount of beneficial drainage. Although the section shown in Fig. 1.4*a* is on a fill, the pavements are poorly drained. Figure 1.4*b* depicts similar designs, but in a cut, with trench drains for control over groundwater. Because of low-permeability materials in the subgrade and shoulder, these pavements are also very poorly drained.

The simplified cross section in Fig. 1.5 shows a typical structural section completely confined within an impermeable subgrade and low-permeability shoulders. The drainability of a trench section like this is determined almost entirely by the environment within 2 or 3 ft of the surface of the pavement, and a structural section can remain in a "bathtub" long after a rain, regardless of whether it is in a cut or on a fill (see Tables 4.2, 4.3, and 4.4).

Figure 1.6 gives typical cross sections through an airfield runway and a taxiway that have no drains and are also in the "bathtub" condition for appreciable amounts of time each year.

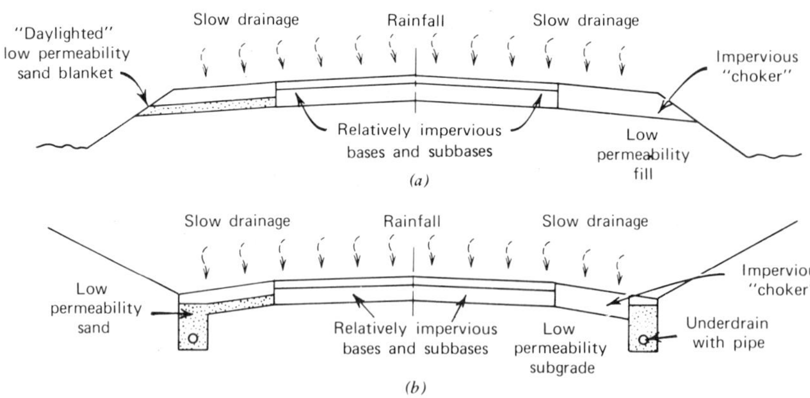


**FIG. 1.4** Conventional modern pavements. (a) Roads on fill; (b) Roads in cut.

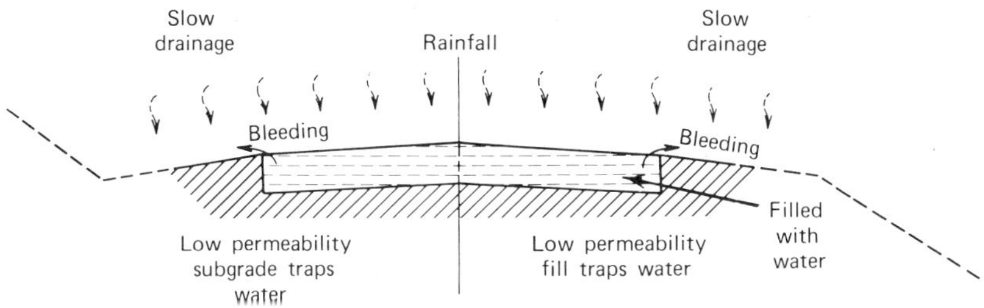


**FIG. 1.5** The immediate environment dominates roadbed drainability. Many pavements are in a "bathtub".

An indication of the general level of the drainage provided by bases and subbases can sometimes be obtained by observing these layers during and after rainstorms that occur during the construction of highways and airfields. Figure 1.7*a*, for example, shows water standing on one of the frequently used types of untreated aggregate base courses several days after a shower. Water still standing in puddles on this layer is an indication of its poor drainability. Likewise, Fig. 1.7*b* shows water still standing on a cement-treated granular base for a partially completed airport runway. Here, also, it is several days since

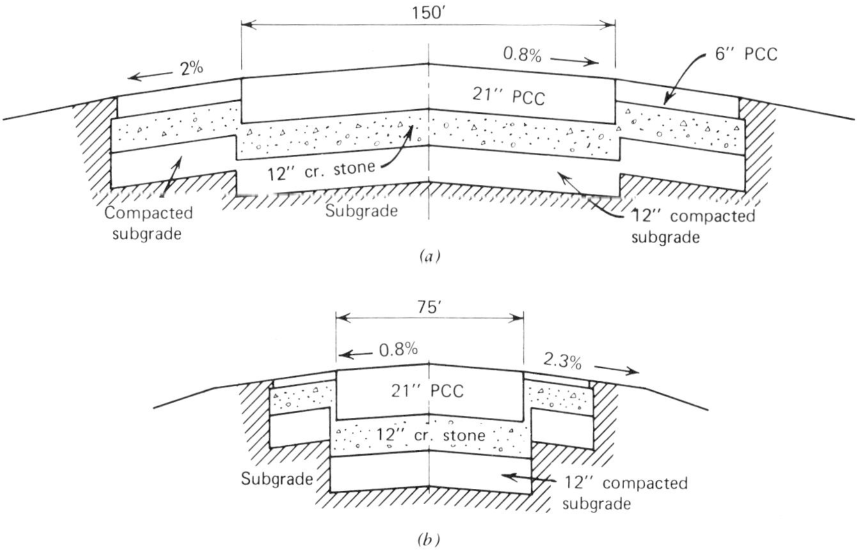


**FIG. 1.6** Typical airfield pavements with no drains. (*a*) Runway. (*b*) Access taxiway.

the last rain, and the water has not yet drained away. Understandably the pavements being built on these supporting layers will not have a high degree of drainability.

Since the idea became popular that pavements can be properly designed without a high degree of drainability (about 1940), there have been no major efforts to provide good structural section drainage in the majority of the pavements that have been built. Consequently it is difficult to find important pavements built within the past several decades (with a few notable exceptions) that have effective subsurface drainage systems primarily or entirely for control of surface water infiltration.

During the field investigation and reconnaissance phase of work that was carried out for the Federal Highway Administration (FHWA, 1973) by the author and KOA (Ken O'Brien and Associates), the interview team inquired in every state visited and on every site-inspection trip if any pavements in that state had been designed with drains to remove surface-water infiltration. Not even one example was called to our attention in any of the 9 states selected for visits, nor did any of the letter inquiries to all 50 states reveal any installations for this purpose. In several of the states visited, however, maintenance forces had put in longitudinal drains along the lower edges of some pavements that were deteriorating from pumping and other damages caused by excess water that entered the structural sections through the tops of the pavements. In all such cases called to our attention, the bases under the pavements were of low permeability; thus only limited benefits were gained by the edge drains.

As may be seen in Fig. 1.8*a*, edge drains that were installed after problems developed have done little to correct bad conditions on this interstate highway. Figure 1.8*b* shows a badly faulted PCC pavement on another interstate highway that has had edge drains installed to try to reduce pumping.

In the field of airport design, edge drains have been used for many years for draining thick, nonfrost-susceptible bases in cold regions (Fig. 1.9), and for control over groundwater, sheet flow, and so forth, but only rarely have they been used to remove surface-water infiltration in temperate climates. Grace, et al. (1973) tell about airfield pavements having severe bleeding problems from slow base drainage and about edge drains that were installed to try to improve the drainage. In many cases little or no improvement was obtained. Regarding one such case they say, "After the subdrains were installed, it was noted that water continued to bleed from the pavement; but no flow was coming from

**FIG. 1.7** The low permeability and poor drainage of these "standard" bases is evident from water still standing several days after rain. (*a*) Highway in an eastern state several days after a shower; September 1970; untreated base. (*b*) Cement-treated base for airport pavement; heavy rainfall area; several days after a rain; 1972.

**FIG. 1.8** Edge drains installed after troubles develop may be relatively ineffective. (*a*) Badly deteriorated AC pavement; edge drain installed by maintenance has not helped; September 1970. (*b*) Badly faulted PCC pavement has been diamond-ground to reduce faulting; edge drain has been put in to try to reduce pumping.

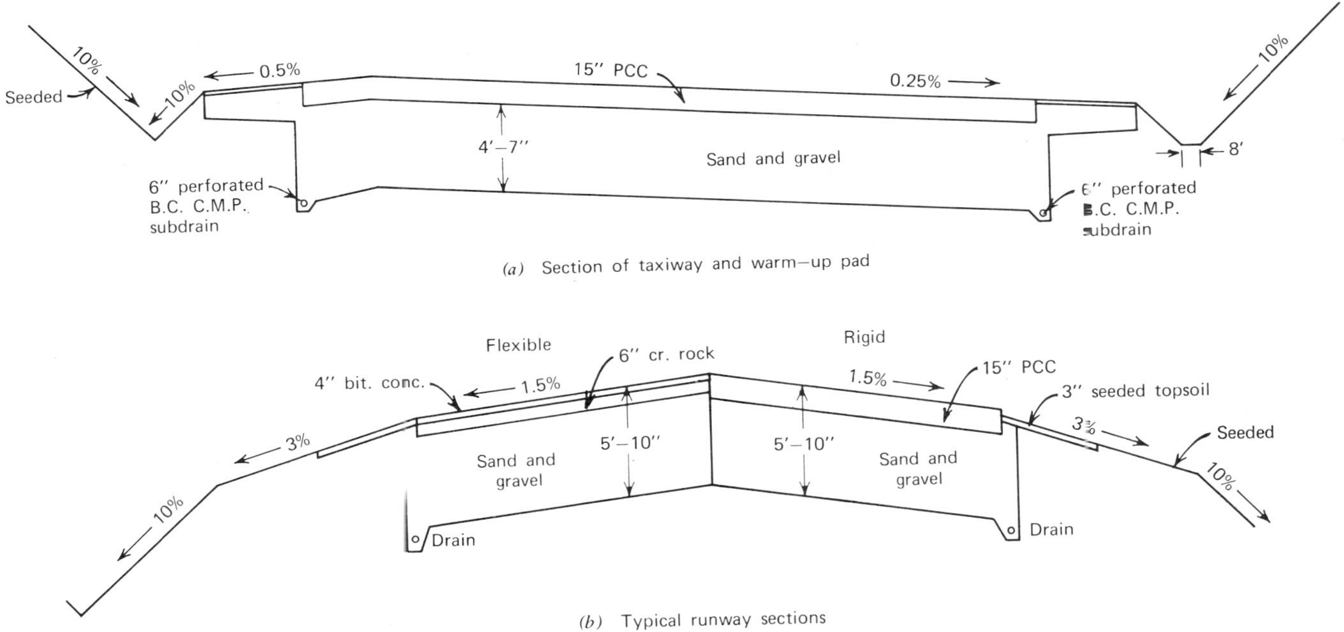


**FIG. 1.9** Airport pavements in cold regions with thick bases and edge drains. (Courtesy CERL.[1])

the system. Several open-end piezometers installed beneath the pavement indicated that the base course continued to be saturated." Where pavements have been constructed on low-permeability bases, it is virtually impossible to do much to improve their drainability after construction.

Theoretically it might be considered possible, without providing for good drainage, to build a pavement sufficiently "stout" to withstand any specified volume and weight of traffic for a specified total number of repetitions or years of service. But, unfortunately, some of the most damaging actions caused by traffic impacts on a section containing free water (pumping, erosion, pulsating pore pressures, etc.) are not significantly reduced by any amount of thickening of base or subbase. As already noted, these actions can occur no matter how thick or stout the supporting base may be; consequently, trying to "live with" excess water can be very uneconomical. The Highway Research Board's Project Committee No. 1, Maintenance of Concrete Pavements as Related to the Pumping Actions of Slabs (Highway Research Board, 1948), says, "The data available show that neither cross section nor thickness have had any effect on pumping. Studies to date have not shown that pavement thickness in excess of that required for imposed loads and the normal supporting value of the subgrade will be either helpful or economically justified."

Spellman (1972) found that heavy wheel impacts on PCC pavements on cement treated bases caused, " . . . violent water actions" at the interface between the PCC pavement and the cement treated base (CTB), causing the erosion of cavities under the PCC pavements, ejection of material out of cracks and joints, and the migration of base material from the leading edges and its buildup under trailing edges. This caused loss of support and also produced the uncomfortable "faulting" or "step-off" that is so common in PCC pavements that are being damaged by traffic and excess water.

In the past few years some investigators have been developing testing procedures with dynamic loading in the hope that this will more closely simulate actual traffic effects. These procedures appear worthy of continued serious investigation.

Major increase in the thickness of the primary pavement layer has been tried as a way to reduce the damage caused by excess water, although it is not a very economical way, and often of doubtful value. Elimination of the water by keeping pavements tightly sealed so far has been of only limited success. *Effective drainage is, of course, another way—the method recommended in this book—and the cases investigated where truly efficient drainage has been employed have been highly successful and economical.*

## 1.3 RECENT PROGRESS IN PAVEMENT DRAINAGE

A comprehensive study that was made for the Federal Highway Administration (FHWA, 1973) to develop a set of *Guidelines for the Design of Subsurface Drainage Systems for Highway Structural Sections* concluded that during the times that free water (primarily from surface infiltration) is contained within structural sections, the rates of damage from heavy wheel loads are undoubtedly much greater in most cases than during the periods of time that no free water is contained in the sections. Reports prepared in that investigation concluded that poor drainage (slow drainability) of heavy-duty pavements throughout most of the United States is a major factor behind the damage that occurs to a majority of these pavements in advance of the assumed design life, and that well-drained pavements will almost always be more economical than poorly drained pavements.

Drainage systems proposed in the FHWA *Guidelines* and in this book make use of a layer of highly permeable open-graded aggregate under the full width of the traveled way, with longitudinal pipes to collect the water and outlet pipes to rapidly remove it from the drainage system.

In the structural design of pavements, coarse crushed rock or gravel has sometimes been used as part of the structural section (see Fig. 6.10*d*). Cases have been examined in which extremely permeable materials of this kind were used, but the potential drainage benefits were not considered necessary, and no outlets or collectors of any kind were provided. As a result, the open-graded bases served as equalizing reservoirs, and water collected at lower elevations (outer edges, sag vertical curves, etc.) where it created excess uplift pressures that led to damage to the pavements. Two entirely independent organizations decided that these coarse, open-graded bases (highly permeable, but with no outlets) were detrimental to the pavements, and they both discontinued their use when they had in their hands the potential for eliminating water damage by the simple procedure of providing collector pipes and outlet pipes.

One state constructed a number of major highway pavements with a thin, dense-graded AC wearing course (¾- to 1-in. thick) over a coarse-graded "binder course" (2.0- to 3.0-in. thick) of very high permeability.

These pavement layers were placed over a stabilized gravel base on an untreated granular subgrade. In some areas maintenance forces allowed the outer edges of the "binder course" to be blocked with asphaltic surface sealing materials, with the result that the open-graded layer became filled with water that could not escape, and the pavements were severely damaged. In locations in which the outer edges were kept free of blockage, this combination served in an outstanding manner.

Notwithstanding the excellent performance of this surfacing combination when it was properly used, it was discontinued because of poor performance at one or more locations in which deficient maintenance practices caused water to become trapped in the open-graded binder course. If pipes had been provided, this could have eliminated the troubles caused by trapped water. Collector pipes and outlet pipes are essential components of subsurface drainage systems, and when properly designed, constructed, and maintained, can assure permanent functioning of these systems in the manner intended at the lowest overall cost.

The conclusions that were reached in cases such as these are difficult to understand. The net result has been that some of the very best kinds of surface and base combinations have been discarded in favor of others that are fundamentally inferior.

An unusual highway project is an "experimental" pavement installed by the Humboldt County Public Works Department (California) on a highly permeable base drainage layer after this section of logging road had failed twice from excess water (Cedergren and Lovering, 1968). The original construction and the prior replacement had been made using "standard" types of construction materials, which were low in permeability. In a high rainfall area, and with sidehill seepage in abundance, the problems were caused by heavy logging trucks pounding pavements filled with water. After the two successive failures using the normal methods and materials, the county decided to try drainage. With the aid of W. R. Lovering of Sacramento, the final reconstruction was designed and was completed in 1967. On the basis of no prior knowledge of how well this reconstructed pavement was performing, it was selected for one of the case study sites for the Federal Highway Administration's study of highway drainage (FHWA, 1973). When it was drilled into after several wet winters, the base and subbase were in " . . . an almost unbelievably well-drained condition," and the pavement was performing exceptionally well.

An essential feature of the Humboldt County experimental pavement was a layer of extremely permeable open-graded asphalt concrete drain material under the primary layers of the structural section. With a coefficient of permeability of 10,000 to 30,000 ft/day, this layer was capable of removing any water that entered within 15 or 20 minutes, which is hundreds to thousands of times faster than can be accomplished with the normally used materials (see Table 4.2). The appearance of this section of road on April 14, 1971, during the field investigation being conducted for the FHWA project, is shown in Fig. 1.10. Although the pavements at both ends of this section were showing some evidence of distress, the experimental well-drained section did not contain a single crack or show any other evidence of difficulties under the traffic of heavily loaded logging trucks that had used the road regularly during the 4 years since its reconstruction in 1967.

Another very noteworthy example of a drainage system, one for a major airport, is the comprehensive subsurface drainage system under the William B. Hartsfield Atlanta International Airport's 8-26 runway (formerly 9L-27R), which was constructed in 1969 after the prior runway pavement had disintegrated under the loads induced by heavy aircraft (Richards, 1971). Although adding only about 5 percent to the cost of the reconstructed pavements, this drainage system will very

**FIG. 1.10** Humboldt County experimental pavement on an open-graded drainage layer of very high permeability (10,000 to 30,000 ft/day); in excellent condition after four wet winters; April 1971.

likely return at least 20 times its extra cost in added pavement serviceability and length of useful life.

An illustration of the potential cost of *not* draining pavements, is a major airfield where $2.5 million has been spent on repairs to its runway system to correct damage that could largely have been prevented by a subsurface drainage system costing about $200,000. After 18 years, 10 times the cost of a drainage system had been spent in repairs, and the problems had not been solved—only temporarily put off.

There can be little doubt that building pavements as slowly draining systems can be a very costly policy in the long run and that any slightly greater initial costs of well-drained pavements will generally be repaid many times over.

## 1.4 SUMMARY COMMENTS

Most of the heavy-duty highway and airfield pavements that have been built in the past 30 years or so are slowly draining systems, largely because standard design practices emphasize density and stability but place little importance on drainage of surface water infiltration. During the periods of time that pavements carry traffic while their structural sections contain free water, the rates of damage can be up to thousands of times greater than when no free water is present. Rising repair and replacement costs of rather new highways and other pavements is considered to be evidence that the design methods are not fully taking into account the effects of traffic impacts on sections containing free water. *If the total long-term costs of important pavements are to be kept to a minimum, design standards must be modified to produce structural sections capable of devoiding themselves of free water rapidly after its entry, rather than in days, weeks, or months.* Rapid drainage of pavement structural sections can be accomplished best by providing a layer of highly permeable, macadam type of open-graded material under the full width of the traveled way, with collector pipes and outlet pipes to ensure constant gravity drainage. Such systems not only protect pavements from large surface inflows, but from inflows from high groundwater in wet periods, spring inflows that do not show up during construction, and virtually any anticipated or unexpected inflows into structural sections (if reserve capacity is provided).

On some occasions highly permeable types of subbases, bases, or binder courses have been placed under pavement wearing courses or bases because of their good structural properties, but they were not thought of or designed as drains, and no outlets were provic consequence, when such practices have been employed, water l been trapped at low places where severe damage to the paveme taken place. Every pavement examined by the author that ha een constructed on a highly permeable, *well-drained base with drainage outlets* was performing in an outstanding manner. There is a good deal of evidence to support the conclusion that properly designed pavements on strong, highly permeable drainage layers can provide long-lasting, high quality service with nearly zero maintenance.

In a report of an investigation to evaluate the effectiveness of drainage and erosion control facilities at 54 airports and air bases throughout the United States, Grace, et al. (1973) concluded that, "... a misguided concept ... appears to have developed in various levels of responsibility in which drainage is considered to be the least important factor affecting the performance of an installation." This concept or attitude (which is surprisingly widespread) is a major factor that must be overcome before substantial progress can be expected in the all-important objective of protecting pavements from the damaging effects of excess water.

Pavements that are built without effective drainage are deteriorating at much faster rates than if they were well drained. Good drainage can save billions of dollars in the long run for those who pick up the tab for pavement construction and upkeep. Furthermore, it can help to conserve diminishing supplies of vital construction materials, since pavements built with well-drained structural sections will use up smaller quantities of these materials over their useful life than those which are not well drained.

## References

Cedergren, H. R. and W. R. Lovering (1968), "The Economics and Practicability of Layered Drains for Roadbeds," *Highway Research Record* No. 215, Highway Research Board.

*Civil Engineering, ASCE* (1970), "Interstate Pavements Failing Early," August 1970, p. 83.

Federal Highway Administration, FHWA (1973), *Guidelines for the Design of Subsurface Drainage Systems for Highway Structural Sections,* prepared by the Joint Venture of H. R. Cedergren, Sacramento, and Ken O'Brien & Associates, Long Beach and Ventura, California, 1972; issued, 1973.

Grace, J. L., Jr., C. C. Calhoun, Jr., and D. N. Brown (1973), "Drainage and Erosion Control Facilities, Field Performance Investigation," Misc. Paper H-73-6, Waterways Experiment Station, U.S. Army Corps of Engineers, Vicksburg, Mississippi, June 1973, pp. 53, xi.

Hewes, Lawrence I. and Clarkson H. Oglesby (1954), *Highway Engineering*, John Wiley and Sons, Inc., New York, pp. 440–1.

Highway Research Board (1948), "Final Report of Project Committee No. 1, Maintenance of Concrete Pavements as Related to the Pumping Action of Slabs," *Proceedings*, Highway Research Board, 28th Annual Meeting, p. 282.

Richards, Robert W. (1971), "Atlanta's Instant Runway," *Transportation Engineering Journal, Proceedings,* ASCE, August 1971, pp. 491–500.

Spellman, Don L. (1972), "Faulting of Concrete Pavements," *Highway Research Record* No. 407, Highway Research Board.

CHAPTER TWO

# Types of Damage Caused by Water in Pavements

## 2.1 HISTORICAL BACKGROUND

Road builders from ancient times have known that water is the greatest enemy of a stable, long-lasting pavement. The ancient Romans, who started building the 50,000-mile Imperial Roman road network in 312 B.C., knew of the damaging effects of water and tried to keep their roads above the level of the surrounding terrain. In addition to constructing these roads with thick sections, they often provided a sand layer on top of the subgrade and below the first course of flat stones that were generally cemented together. The durability of those highways is proved by the fact that many of them still exist.

Little basic progress was made during the 20 centuries after the road building of the Roman Empire, until Tresaguet, Metcalf, Telford, and McAdam in the first half of the nineteenth century introduced important improvements in the then-prevailing road building methods (O'Flaherty, 1967). One of the most important principles they "rediscovered" was the need of keeping roadbeds dry. In 1820 John L. McAdam made his oft-quoted statement to the London Board of Agriculture (McAdam, 1820) that,

> The roads can never be rendered thus perfectly secure until the following principles be fully *understood, admitted,* and *acted upon*: namely that it is the native soil which really supports the weight of traffic; that whilst it is preserved in a dry state it will carry any weight without sinking . . . that if water pass through a road and fill the native soil, the road *whatever may be its thickness* loses support and goes to pieces.

He goes on to say,

> The erroneous opinion so long acted upon, and so tenaciously adhered to, that (by) placing a large quantity of stone under the roads, a remedy will be found

for the sinking into wet clay, or other soft soils, or in other words, that a road may be made sufficiently strong, *artificially*, to carry heavy carriages though the sub-soil be in a wet state and by any such means to avert the inconvenience of the natural soil receiving water from rain, or other causes, *has produced most of the defects of the roads of Great Britain.*

Later in this chapter somewhat detailed descriptions are given of the kinds of damage that are caused by the dynamic pounding of heavy wheel loads on pavements in which free water fills the pore spaces and is trapped between the various layers of pavements, bases, and subbases. It will be brought out that modern design methods are not fully taking into account these dynamic actions. Ironically, they could not take place at all if McAdam's advice were being followed, and there were no free water to be acted on. These damaging actions are not being stopped by the use of artificial methods for strengthening pavements. Thus, McAdam's complaint (made in 1820) is just as valid today, more than 150 years later.

Although the significance of good roadbed drainage has been acclaimed on and off through the centuries, it has often been unheeded by road designers, and little attention was given to drainage of roadbeds during the latter half of the nineteenth century. In 1910, however, the merits of drainage were clearly brought out in *The Art of Roadmaking* (H. Frost, 1910) in statements such as, "A road on a wet, undrained bottom will always be troublesome and expensive to maintain, and it will be economical in the long run to go to considerable expense in making the drainage of the subsoil as perfect as possible."

From about 1910 to about 1940, pavement engineers widely proclaimed the need for drainage of roadbeds (and railroad beds), and nearly every textbook on road building or railroad construction and many published articles contained a statement to the effect that, "There are just three things that are necessary to get a good roadbed, and they are *drainage, drainage,* and *more drainage.*" But, with the advent of the rational methods for designing pavements and laboratory testing of saturated samples of subgrade and base course materials, the emphasis moved to density and stability rather than drainability. The practice has been to try to design sufficiently "stout" by the use of sufficiently thick erosion-resistant bases and subbases and stabilized subgrades, so that pavements can carry the required loads in the presence of water. Also, there has probably been a tendency to underestimate the amount of water that is available to enter most pavements—particularly from natural precipitation.

Leading up to the practice of designing pavements for saturated strengths were many detailed studies of moisture conditions under pavements, which disclosed that even in semi-arid and arid climates, water can accumulate under pavements. Myers et al. (1941) describe the effects of many factors (capillary phenomena, humidity, evaporation, frost, and hydrodynamic effects) on soil moisture phenomena under pavements.

Kirsten (1944) summarizes a detailed survey of moisture conditions under pavements in several states, which disclosed that a substantial proportion of all of the subgrade types tested had moisture contents in excess of the optimum. Kirsten (1946), in summarizing a study of moisture contents of airport pavements, reports that moisture conditions varied with the soil texture and climate of a region, with higher saturation in heavier textured soils, but less water under pavements in arid to semi-arid regions than in heavy rainfall regions.

One of the first design procedures using saturated strengths of subgrades and bases was the CBR test which was originated in California by O. J. Porter. Porter (1938) explains the necessity for soaking test specimens, as follows:

> Test results indicate that clay, adobes, and other adverse soils usually have a good supporting value when in a thoroughly compacted and relatively dry state. This ideal condition, however, does not maintain over a period of years. Such material usually absorbs sufficient moisture from rainfall, ground water or by capillarity, to cause expansion and, as a consequence of the increase in water content, the soil often reaches a state of compaction and wetness comparable with the soaked specimen in our bearing value test.

When the design of a pavement is based on the testing of saturated specimens, there is a general expectation that there is no need to make further allowances for environmental factors such as the amount of rain that falls on the pavement. The modern design techniques generally assure sufficient thickness of pavements and bases to prevent overstressing of subgrades; however, as already noted, *some of the most serious damage from water will occur regardless of the thickness or stability of base courses used*, because it is caused by pore pressures and movements of free water contained in structural sections.

Even though most of the important pavements of the past three decades have been designed on the premise that surface water is *not* important, some investigators have still expressed the belief that environmental factors such as rainfall *are* important and should be taken into consideration in designing pavements. For example, Finn (1962) says he believes that work should be done to develop an environmental adjustment factor since critical periods of pavement performance might be associated with rainfall. He recommends that application of adjustments," . . . should be introduced to design equations only after extensive research."

The Highway Research Board's Committee on Rigid Pavement Design (HRB, 1968), says that there is a need to evaluate the effects of environment, including climatic and regional variables (moisture and temperature) on pavement on a nationwide basis. Finney (1973) says,

> The engineer with a knowledge of local environmental factors can use the information from the subgrade surveys and classification tests to estimate the susceptibility of the subgrade to volume change, frost action, and pumping. Also, he will have a certain knowledge as to the supporting properties of soil under load, and then he can make design recommendations accordingly.

## 2.2 THE MECHANISMS OF DAMAGE TO PAVEMENTS

### Basic Factors

The whole subject of distress and failure of pavements is very complex, because many factors enter into pavement deterioration and failure. The aging and oxidation of asphalt films enters into the deterioration of asphalt concrete pavements and bases. Certain deterioration processes and fatigue under repeated strain are major factors in the cracking up and disintegration of PCC pavements. Some of the most detrimental actions occur more rapidly when excess water is in the spaces between successive lifts of AC pavements or in pore spaces in these pavements, in joints and cracks in PCC pavements, or at the boundaries between PCC pavements and their bases.

Pavements have large surface areas exposed to weather and traffic. Under the combined actions of temperature changes, rainfall, freezing and thawing, oxidation, and so forth and the stresses and strains of traffic impacts, a number of kinds of damaging actions take place that shorten life, increase roughness, put loosened material on surfaces, and eventually destroy many pavements. A considerable amount of research and many special studies and investigations have been conducted on the mechanisms of pavement deterioration and failure by a large number of workers and agencies, and many reports and papers have been published. If one reviews the available literature and reports and the detailed studies that have been made by the author, it is evident that water is a leading factor in causing damage to pavements and that high repair and replacement costs on major highways and airfields are nearly always associated with heavy wheel loads and impacts on sections containing free water.

Several photographs in Fig. 2.1 show pavements that have an abundance of water available to filter through cracks, joints, and so forth into the structural sections. Figure 2.1*a* shows a portion of a field in a semiarid climate in the west during one of the days. Water can be seen standing on the surface after a sh quite apparent that at times this pavement is well inundated. Figure 2.1*b* shows a secondary state highway in a Pacific coast area of a western state. This area receives at least 40 in. of rainfall a year, plus a good deal of moisture from melting frost (no deep freezing) and condensing mist, as at the time of the photograph. This road is in a rather advanced state of deterioration. Figure 2.1*c* shows a "spring" flowing out of an airfield runway in a New England state several days after a rain. General bleeding of water out of this area had been going on for years after rains. The hole in the foreground, chopped in the pavement the day before the photograph, has allowed a large amount of water to flow out of this pavement from inflows that entered at higher elevations. This pavement is in the "bathtub" condition and is even exposed to excess uplift pressures many months out of each year. Figure 2.1*d* shows a badly cracked asphalt concrete pavement on a state highway in an eastern state. There is still some water in evidence, although this pavement is on a shallow fill with no groundwater seepage, and it has not rained for several days. There can be little doubt that many pavements are required to carry loads while their structural section is in a supersaturated condition.

Some of the most harmful actions of traffic and water occur entirely within the structural section of a pavement and are unrelated to subgrade strength or behavior. Adams (1969) tells about investigations of distress of flexible pavements by the State Highway Department of Georgia. Three locations on interstate highways were investigated because of evident distress, and none of the three showed any evidence of subgrade distress. Likewise, the case study investigations that were made as part of the field work for the FHWA *Guidelines* project (FHWA, 1973a; FHWA, 1973b) showed serious pavement distress at eight interstate or state highway test sites, but there was no evidence that subgrade weakness was a factor of major significance.

As the water content of bases and subbases increases, there is a reduction in supporting power and an increase in the rate of loss of serviceability of the pavements. When free water completely fills these layers and fills the voids and spaces or openings at the boundaries between layers, heavy wheel loads applied to the surfaces of these pavements produce impacts on the water that are comparable to a water-hammer type of action. The pulsating water pressures that can build up at such times under wheel impacts not only cause erosion and ejection of material out of pavements, but can strip asphalt coatings from bituminous-stabilized bases and subbases. The water actions can

**FIG. 2.1** Surface water infiltration is often an abundant source of water in pavement structural sections. (*a*) Water ponding on airfield pavement after light rain; in semiarid climate in western state; December 1972. (Courtesy CERL.[1]) (*b*) Condensed dew and melting frost; secondary state highway in west coast area; no deep freezing; December 1968. (*c*) "Spring" flowing out of airfield pavement several days after it rained; hole was dug in pavement; New England state; August 1972. (Courtesy CERL.[1]) (*d*) Eight-year-old AC pavement; badly damaged from traffic and excess water; no groundwater; several days since it rained; eastern state; September 1970.

(c)

(d)

disintegrate cement-treated bases, weaken base courses by rearranging the internal structure of fine-grained materials in aggregate mixtures, overstress subgrades where total thicknesses are inadequate, and cause a number of other detrimental actions.

A special Highway Research Board report (HRB, 1952) describes the sequence of damage to concrete pavements containing free water as follows:

> The first step in the development of pumping is the breaking of the seals of adjoining slab ends and the loss of intimate contact between the free edges of the pavement and the shoulders, which of course, allows surface water to reach the subgrade soil. Seals at transverse joints and open cracks are broken by contraction of the adjoining slabs as the temperature of the pavement drops, by relative deflection (deflection of the slab on one side of the joint or crack with respect to that on the other) as vehicles pass over the pavement and by other factors, such as weathering and poor initial bonding. The loss of intimate contact between the shoulders and the edges of the pavement is caused by (1) shrinkage of the shoulder material, (2) excessive deflections of the slab edges, (3) rutting of the shoulders and (4) in pumping pavements, the ejection of water from the subgrade to the surface.
>
> A vehicle in moving over the pavement develops pressure between the bottom of the slab and the subgrade which causes water, if present on the subgrade, to surge or flow in the direction in which it might escape. The magnitude of the pressure is, of course, dependent upon the deflection of the pavement slab and the degree of confinement of the water. The water may, in the beginning, escape only through joints or open cracks, but in time, if susceptible pumping conditions obtain, it generally develops an avenue of escape between the shoulder and the edges of the pavement. In pavements constructed on a fine-grained subgrade soil this surging and flow of water beneath the slab causes fines in the soil to mix with the water, which, in escaping, carries the fines with it. A void, becoming progressively larger with repetition of traffic, then forms beneath the pavement slab.
>
> In the early stages of pumping the escape of the water or slurry between the shoulder and the edge of the pavement is, as a rule, confined within several feet of a transverse joint or open crack; but as pumping progresses, an escape channel frequently develops along the full length of the pavement slab. As edge pumping increases a greater volume of water is permitted to reach the subgrade thus increasing the amount and the intensity of pumping.

With free water completely filling pavements, *buoyancy* of supersaturated layers reduces unit weights to submerged weights, thereby reducing frictional strengths (usually a secondary factor). As already noted, free water provides the medium for the erosion of material at boundaries between pavements and bases (both untreated and stabilized) and for the physical ejection of material by "pumping" out of cracks and joints, leading to faulting or step-off and ultimately to total failure of PCC pavements, and serious damage to AC pavements.

Free water in AC pavements contributes to shrinkage cracking and to oxidation and loss of flexibility, which can lead to cracking and general deterioration of wearing courses and stabilized bases.

Free water serves as the medium for clay or other fines to be forced into incipient cracks in the under sides of AC pavements, preventing the self-healing of these cracks.

The presence of free water allows *frost action* to occur in susceptible materials, and the prolonged exposure to free water has been associated with "D" cracking and "blow-up" of PCC pavements.

Free water is almost always associated with the breaking out of chunks of pavements constructed of two or more successive layers of similar paving material.

It would be difficult to single out any one of the foregoing mechanisms and actions as being the most damaging to pavements. In some areas, *frost action* is an extremely damaging problem. In others, D cracking is of grave concern. But, in nearly every area of our country, the effects of excess pore pressure and erosional actions are among the most severe causes of loss of pavement life.

In the studies for development of the *Guidelines for the Design of Subsurface Drainage Systems for Highway Structural Sections*, (FHWA, 1973b pp. 46–69), nine case study sites were diamond drilled; test pits and auger holes were put down; and detailed examinations were made for any damage caused by water. One of these test sites, located on a 50-ft-high fill, had 12-year-old interstate highway pavements that were showing severe damage from excess water. As with the other interstate and state highway sites, which were all on fills, the supply of water had to be from precipitation because no groundwater or other seepage could get into the roadbed. The core drilling at this site revealed another kind of deteriorating action—the complete loss of cohesion of the cement treated base material under the PCC pavement. A hole that was cored into the interior of a slab *in the truck lane and directly under the wheel path*, disclosed that the CTB had deteriorated into a completely cohesionless material that could be removed with a spoon or with the fingers. A maintenance foreman told the investigating engineer that this is a common occurrence along this highway, and that at every location where he had dug out sections of failed pavement to be replaced with high-early strength concrete patches, the CTB was soft and disintegrated, as at this site.

Another hole was then cored into the passing lane at this site, and a very solid core of cement-treated base was recovered along with the concrete (see Fig. 2.2), indicating that the loss in cohesion in the truck lane was undoubtedly caused by the combined actions of traffic and water. As a result of these investigations, the conclusion was reached that pulsating pore pressures under heavy wheel impacts must cause some kind of disintegrating action—possibly by the miniscule surging of water under pressure into and out of the cement treated base. Individual water movements are undoubtedly of microdimensions, but the repeated actions may be causing a leaching or another effect on the cement, leading to a complete loss of cementing. For this kind of action to occur the pores and spaces in the structural section must be filled with free water, in which pore pressure waves move along with the heavy wheel loads. Figures 1.2*a* and *b* show evidence of the "bathtub" condition and excess water at this site.

In surveys of problems with both highway and airfield pavements, D cracking was of major concern in a number of northcentral and New England states, which have low-quality aggregates, high rainfall, and extensive freezing and thawing, all of which contribute to D cracking. Figure 2.3*a* is an example of the typical pattern of D cracking at the intersections of thick PCC pavement slabs at several airfields. Figure 2.3*b* shows a spalled D cracking area on an airfield that had been temporarily patched with asphaltic material. Figure 2.3*c*

**FIG. 2.2** Core samples of PCC and intact CTB; passing lane. In the truck lane (outside lane) the CTB was completely disintegrated from surface water infiltration and traffic; on 50-ft fill, no groundwater; interstate highway in western state; December 1970.

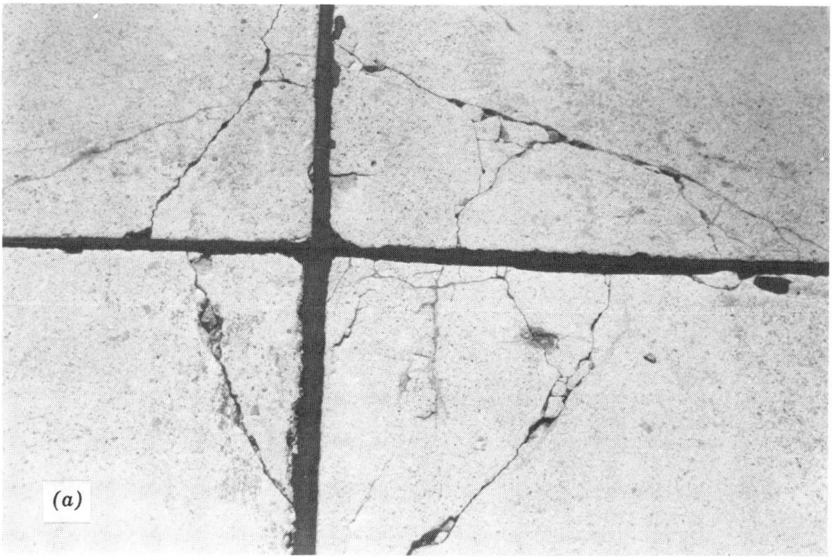


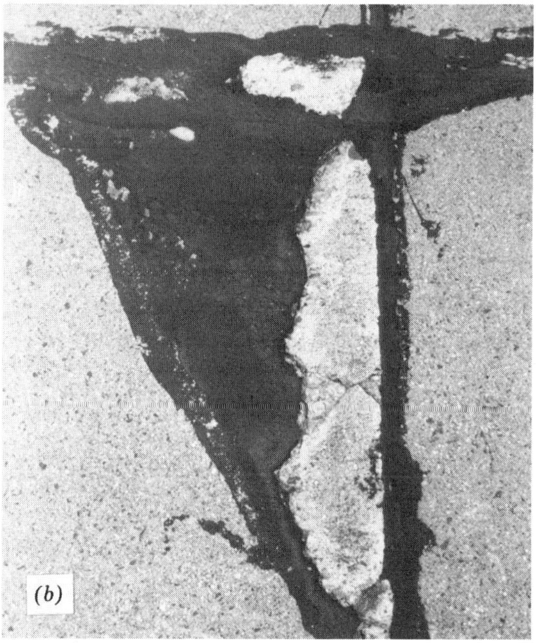


**FIG. 2.3** Types of pavement deterioration (PCC) that may be increased by slow drainage. (*a*) D cracking on airfield taxiway; Great Lakes area; slow drainage, and much rainwater; August 1972. (Courtesy CERL.[1]) (*b*) D cracking and spalling of airfield apron pavement; New England area; August 1972. (Courtesy CERL.[1]) (*c*) D cracking near open edge joint of an interstate highway in a central state; probably aggravated by slow drainage; September 1970.

**FIG. 2.3** (Continued).

is a photograph of a D-cracking popout condition near an open edge joint in an interstate highway in a central state, which has an abundance of rainfall and much freezing and thawing.

An excellent report of D cracking of concrete pavements by the Portland Cement Association (Verbeck et al., 1972), in telling about detailed surveys of some 4400 lane miles of pavements in Ohio, describes the D cracking process as follows: "The field and laboratory observations noted above are taken as evidence that D cracking is caused by stresses generated during the freezing of critically saturated coarse aggregate particles." Also, it says,

> Deterioration of this type is initiated when atmospheric moisture penetrates open joints and cracks, and together with moisture already present beneath the

pavement, raises the degree of saturation of the coarse aggregate to a critical level. During freezing, pressures generated in the aggregate through hydraulic pressure or ice formation then cause disruption of the aggregate and surrounding mortar. With continued freezing and thawing, existing cracks may provide additional channels for migration of moisture into the slab and also become additional sites for ice formation and the generation of excessive pressures operating to widen the existing cracks. If allowed to progress, the entire pavement slab will be converted to an incoherent mass of rubble.

Under "Conclusions and Recommendations," the PCA report says, "Laboratory studies of pavement cores from selected locations on I-71 revealed the importance of the availability of moisture to the pavement slab and the nature of the coarse aggregate in the development of D-cracking." And it goes on to say,

Somewhat lower degrees of saturation were found in cores taken five or more feet away from the joint areas, which suggests that moisture which contributes to the saturation level of the concrete indeed filters through open joints. It is thus recommended that precautions be taken to upgrade the coarse aggregate and reduce the moisture flow through joints.

Since it is virtually impossible to prevent moisture flow through joints and cracks, it appears to this author that the rapid elimination of water by good drainage offers a practical, economical way to reduce D-cracking damage.

Since some of the damaging actions in pavements (HRB, 1952; Spellman, 1972) are caused by the velocity of water being forced about in structural sections, an indication of the probable rates of damage with age can be obtained by looking at the fundamental equations of flow of water in enclosed channels, pipes, etc. According to fundamental hydraulics, velocities of flow vary with the square root of pipe or channel diameter. Therefore, when a pavement is new and channel dimensions are small, velocities of the trapped water will be smaller than after the channels have been enlarged and extended by the erosion. Consequently, as a pavement gets older and the erosion channels and spaces get larger, the rate of movement and quantity of water moved and the *rate of erosion* can be expected to increase geometrically with time and traffic. Other deteriorating factors such as fatigue, oxidation, and increased volume of traffic add to the propensity for accelerated rates of damage with age. As a result, a pavement might appear to be functioning fairly well at the end of 7 or 8 years, and be severely damaged at the end of 15 or 16 years, for example.

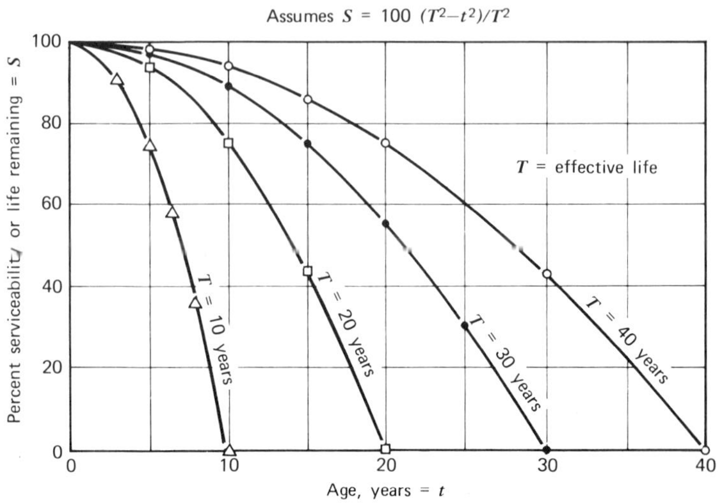


**FIG. 2.4** Chart showing hypothetical rates of losses in serviceability of pavements of various amounts of effective life (assumes that rates of deterioration increase with *square* of time.

Figure 2.4 is an illustration of the way the serviceability of many pavements might diminish with time. Curves of percent serviceability or life remaining versus age in years are given for pavements having an actual effective life of 10, 20, 30, or 40 years. These curves were developed on the assumption that the combined effect of all damaging actions is reducing serviceability and cutting pavement life at rates that increase with the square of time. For this assumption, the remaining serviceability $S$ can be estimated from the following expression:

$$S(\%) \sim \frac{100\ (T^2 - t^2)}{T^2} \tag{2.1}$$

In Eq. 2.1, $T$ is the actual effective life in years and $t$ is the age at a given time for which $S$ is desired.

By referring to Fig. 2.4 it is seen that if a pavement is to have an effective life of 20 years, it should still have 75 percent of its serviceability left after 10 years (for this assumption). But if this pavement is being damaged at a faster rate than allowed for by the design formula, and if its serviceability is lowered 25 percent in the first 5 years, its actual effective life might be only 10 years. On the other hand, if the rate of damage is significantly reduced, for example, through good drainage, its effective life might be lengthened to 30 or 40 years. If its effective life is 30 years, its serviceability should be reduced only about 12 percent at the end of 10 years, if its effective life is 40 years, its serviceability should be reduced only about 6 percent at an age of 10 years.

If it were possible by examination of pavement condition to estimate the present serviceability of a pavement, charts such as Fig. 2.4 would provide a way to roughly estimate the number of years of life remaining (if the assumptions made in developing the charts are reasonable). Any such estimates should be considered rough approximations, because many things can change in the life of a pavement to either accelerate or slow down the rate of damage.

### Importance of Environmental Factors

The significance of seasonal environmental factors such as rainfall and the spring thaw on rates of deterioration of pavements is well known to the maintenance forces who have been keeping pavements in operating condition throughout the past, for these people have been kept very busy during wet weather and during the spring "break-up"—both of these periods being times when large amounts of water are available to infiltrate or otherwise accumulate within structural sections. Likewise, the importance of environmental factors has been well established from detailed records kept of a number of road tests performed since 1950. Among the well-documented tests providing factual information about environmental conditions and rates of loss in serviceability are Road Test One-MD (in 1950), the WASHO Road Test (in 1954), and the AASHO Road Test (in 1958–1960). These road tests provide records of much faster losses in serviceability during unfavorable environmental conditions than during more favorable times (see Sec. 2.4, "Excess Water and Shortened Pavement Life").

Even though all of these road tests had much greater rates of pavement damage during wet periods than during dry, this has not led to any significant change in pavement design practices beyond the use of stronger pavements and "erosion-resistant" types of bases. Until the issuance of the FHWA *Guidelines* in 1973 (FHWA, 1973a), there had been no coordinated effort in several decades to improve drainability and reduce exposure of structural sections to excess water.

Detailed studies for preparation of the FHWA *Guidelines*, which included exhaustive discussions with highway engineers in nine selected states and examination of hundreds of miles of "typical" pavements from coast to coast, led to a number of conclusions such as the following, which are contained in various reports, including the "Final Report," prepared in that study (FHWA, 1973b):

(1) Water can get into most pavements through joints, cracks, porous surfaces, high shoulders, and so on, since the present state of the art does not assure watertightness of pavements for long periods of time. Thermal changes, traffic, and other environmental changes and effects are continuously working to develop openings in pavements.

(2) The rates of damage to most pavements are much greater when excess water is in structural sections than when there is little or no excess water there.

(3) Rainfall rates in most parts of the United States are greater than coefficients of permeability of subgrades (natural or compacted); hence the well-known "bathtub" condition tends to exist in most pavements for significant amounts of time each year.

(4) As a consequence of items 2 and 3, most modern roads, airfields, and other important pavements are actually very slowly draining systems, and many contain excess water for extensive periods of time a number of times each year, which greatly accelerates losses in serviceability and increases repair and replacement costs.

Various other investigations that have been made have pointed up the importance of surface water infiltration in pavement deterioration. Spellman (1972), in discussing detailed field investigations of the faulting of 3- to 8-year-old interstate pavements in California, with PCC wearing courses on cement-treated bases, tells about finding cavities under leading slabs causing distinct faulting or step-off at cross joints. Traffic impacts in wet climatic periods were said to be responsible for high pressures in the trapped water, causing strong jet-like actions of the water that eroded even the cement-treated bases, ejected material out of cross joints and edge joints, and led to the buildup of material under the trailing slabs, further increasing the amount of the faulting. The pavements investigated in Spellman's study were constructed in the 1960s using the latest designs in effect at that time.

Figure 2.5 shows the way concrete deterioration can progress when faulting is taking place. In Fig. 2.5*a*, a 5-year-old pavement on an interstate highway in a north central state has developed approximately 0.4-in. faulting at cross joints, although the pavement itself does not yet show any structural failure. In Fig. 2.5*b* a more advanced stage of deterioration of a concrete pavement in a New England state has developed after 8 years of heavy traffic. Both of these pavements are on fills; hence the damage is the result of traffic impacts on pavements containing free water that has entered through surface

**FIG. 2.5** Deterioration of PCC pavements is a gradual process. (*a*) Five-year-old interstate highway in a north central state; 0.4-in. faulting but no cracking; September 1970. (*b*) Eight-year-old interstate highway in New England state; badly broken pavement on 20-ft-high fill; no groundwater; October 1970.

cracks, joints, and the like, because no groundwater can get into either of these pavements.

The Highway Research Board's Special Report 4 (HRB, 1952) describes the traffic test procedures and summarizes the results of Road Test One-MD, an extremely thorough and well-documented experimental road test on an existing, 10-year-old PCC pavement built directly on the local Maryland soil subgrade. Two adjacent 12-ft-wide lanes of 9 in.-7 in.-9 in. reinforced PCC were tested with concentrated traffic of both single- and tandem-axle trucks from June through December 1950. During the 6 months of the test operation, a total precipitation of 23.48 in. was recorded. Pumping, cracking, and deterioration of pavements progressed throughout the test; however, toward the end of the period, as the weather became colder, pumping and the rate of damage sharply increased. According to the HRB report,

> The climatic conditions, temperature and rainfall, have an effect on pavement performance, since (1) shrinkage of subgrade soils and warping of pavement slabs caused by change in temperature may result in loss of subgrade contact and therefore assist in the movement of rain water between the bottom of the slab and the top of the subgrade, and (2) it requires free water under the deflecting pavement slab to develop the soil-water slurry (mud) which is ejected by pumping action of the slab under traffic.

The report also says, "All other factors being equal, the evaporation rate increases with the increase in temperature; thus pavement pumping after a rain usually continues for shorter periods in hot weather than in colder weather."

A comparison of amounts of pumping of joints in cut and fill sections in the Maryland test shows no significant difference, indicating that the immediate environment often overshadows broader factors such as the position of pavements relative to the elevation of the adjacent terrain (see Fig. 1.5).

In an analysis of the structural effects of heavy-vehicle loads, the HRB Special Report 4 states:

> There is no practical, economical cure for pumping in a concrete pavement which can be depended upon to be permanent once it has developed to a marked degree, and as shown by data previously presented, the stresses which caused structural failures at pumping joints were influenced more by the development of pumping than by actual stress differences in the vehicle loads used in the traffic tests.

The HRB report further points out that prior to termination of the traffic test, essentially all the joints in the slabs on the clay subgrade were pumping, and nearly half the slabs were severely cracked. At the end of the test, losses of material by pumping had caused some pavements to settle nearly 4 in., and very severe breakup of many slabs had occurred.

The Highway Research Board's National Advisory Committee (HRB, 1962) reports on the performance of the AASHO road test pavements at Ottawa, Illinois, 10 years after Road Test One-MD. Reports of this committee pointed out that none of the AASHO pavements was provided with an underdrain, that the shoulder materials were relatively impermeable aggregate mixtures, and that the subgrade was very low in permeability; hence these pavements were in a "bathtub" during wet periods. Typical cross sections through both rigid and flexible types of pavements in this road test are given in Fig. 2.6. This report says, ". . . free water that collected under the slab during rains did not drain laterally through the subbase material in the shoulder at a rate sufficient to prevent pumping."

Studies carried out in the AASHO road test indicated that all failures of the rigid pavements were preceded by pumping of material from under concrete slabs. The report recommends additional research on PCC pavements constructed on erosion-resistant bases.

### Possible Remedies

Extensive field tests and observations have indicated that rates of damage and loss in serviceability of both rigid and flexible types of pavements are much greater when structural sections contain free water than when they do not. Strassenmeyer and Wilson (1967) review the results of a study of edge pumping of PCC pavements in Connecticut. A number of sections of Interstate 91 showed evidence of edge pumping shortly after it was opened to traffic. To remedy such problems, these writers recommend that better efforts be made to seal joints, stricter materials control be practiced, and *underdrains that are capable of providing effective control over the water that gets under the pavements* be utilized.

Several of the state and interstate highway case study sites tested in the development of the FHWA *Guidelines* (FHWA, 1973b) had AC or PCC pavements 6 to 8 years old that were prematurely damaged from the effects of traffic and excess water. The entrapment of surface water in the structural sections of these pavements by low-permeability subgrades and impermeable shoulders and its slow drainage after infil-

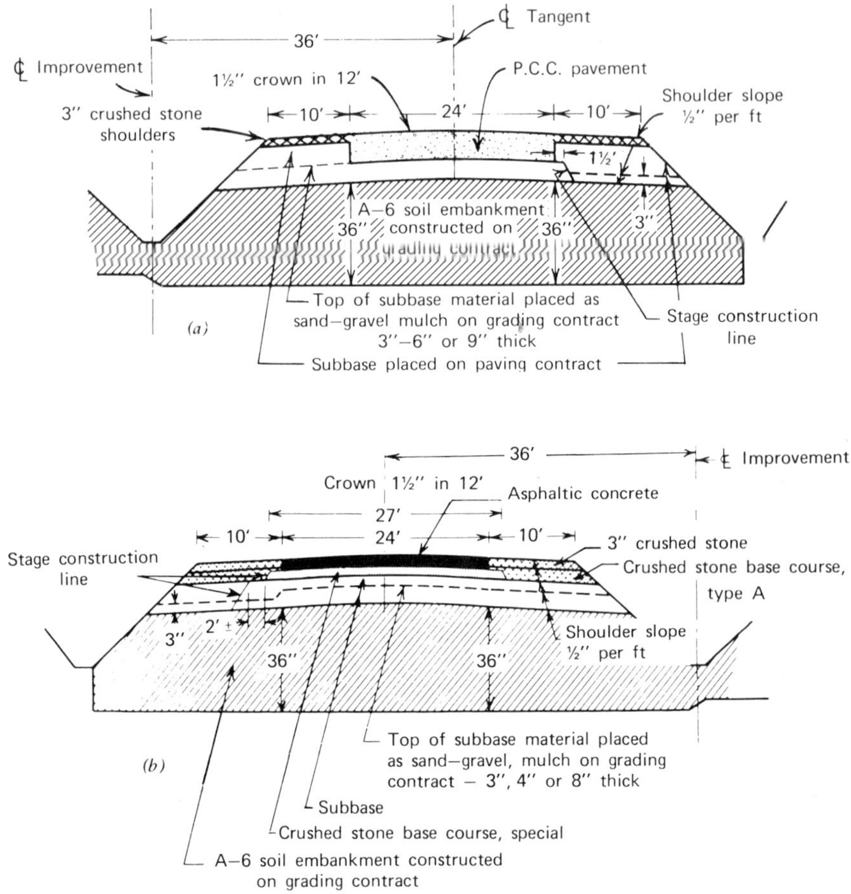


Note: Table 1 of the HRB report gives thickness of subbase, base, and asphaltic concrete included in each loop.

**FIG. 2.6** Typical cross sections through rigid and flexible pavements on AASHO road test. (HRB, 1962).

tration were reported to be major factors in the damage to these pavements.

Changing designs of pavements to try to reduce the damage caused by traffic on structural sections containing excess water *without increasing drainability* does not remove the basic *cause* of deterioration; hence, this is a costly method and one using up large amounts of essential materials in relation to the benefits obtained. In some states PCC pavements are extensively reinforced with steel to try to prevent or reduce faulting and pumping. In some areas continuously reinforced PCC pavements are said to be performing well; in others the results of reinforcing efforts are said to be quite ineffective. In the design of AC pavements, some designers are advocating stabilization of the entire depths of these pavements, their bases, and subbases.

On the whole, efforts to change details without improving drainability are having only little effect on the amount of deterioration and failure in relation to cost. McCullough (1970), in presenting criteria for designing reinforced concrete pavements, cautions that, "Continuously Reinforced Concrete Pavement may be susceptible to edge pumping if a poor subbase material with an excessive amount of fines is used, or if *heavy traffic and excess water are present.*" This combination of conditions is, of course, harmful to both rigid and flexible types of pavements.

A multitude of experience and evidence points up the great potential advantage of draining important pavements, rather than trying to make pavements sufficiently strong, *artificially*, to carry heavy wheel loads without drainage. The studies made in developing the FHWA *Guidelines* previously noted (FHWA, 1973b) led to the following basic conclusions regarding the prevention of water-induced damage to pavements:

(1) When all costs over the useful life of an important pavement are considered, a well-drained pavement will almost always be more economical than the same pavement designed as a slowly draining structure.

(2) The design of a subsurface drainage system should be on a rational basis using seepage principles to calculate probable inflow quantities and to determine required outflow conductivity.

(3) If a pavement is to be drained fast enough to significantly reduce water-induced damage, a highly permeable macadam type of base drainage layer is needed under the full width of the heavy traffic lanes, fitted with longitudinal collector pipes and outlet pipes to ensure free gravity drainage out of the system and thus prevent harmful accumulations of water within the structural section.

(4) The special base drainage layer needed for rapid drainage of a roadbed generally can be substituted on an inch-for-inch basis for the nondraining bases and subbases being replaced by the drainage layer. Frequently the only added cost for a well-drained highway or airfield pavement will be the cost of the collector pipes and outlet pipes.

(5) The fast-draining layer should be constructed of strong, durable aggregates, generally in the range of ¼- to 1-in. particle size. No particles smaller than about ¼ in. should be allowed in any base drainage

layer because smaller sizes can seriously cut permeability, and the upper size should be kept within the range of ½ up to a maximum of 1½ in. (depending on the conductivity needed in individual cases).

(6) Crushed durable rock or gravel with high frictional strength and high interlocking properties will often be sufficiently stable without the use of additives or binders; however, if the aggregates are largely rounded, or if added cohesion is needed for the structural requirements, construction stability, or permanent stability, it may be necessary to stabilize these materials by plant-mixing them with 2 to 4 percent hot paving-grade asphalt or other suitable binders.

A leading civil engineer has said that well-drained structures are nearly always more economical and efficient than their undrained counterparts. A good deal of evidence suggests this is just as true of pavements as of other engineering structures that are subjected to the influences of free water. In an era of diminishing supplies of natural resources, techniques that can conserve vital construction materials will need to be seriously considered.

## 2.3 PHYSICAL EVIDENCE OF EXCESS WATER IN PAVEMENTS

Evidence of pumping and faulting often can be seen when PCC pavements are very new, even though the pavements otherwise appear in perfect condition. Thus Fig. 2.7 shows bleeding and staining on the shoulder of a 1-year-old PCC pavement on a cement-treated base (CTB). Although the design was the lastest in use in this state, it is a slow-draining pavement, and ⅛ in. of faulting at cross joints has already occurred. The pavement is in good structural condition, although it might be predicted that more evident problems will show up in advance of its design age.

Because of the effects of excess water and other compounding factors such as fatigue of concrete under repeated stress, many pavements that appear to be in excellent condition after several years of service, may after a few more years rather suddenly require extensive mud jacking, grinding, and other repairs. When evidence of pumping and loss of material can be seen in new pavements, as in Fig. 2.7, extensive repairs or replacements will probably be required long before they reach their design ages. Certainly it is evident that these pavements are having to carry loads while in the "bathtub" condition previously noted.

Asphalt concrete pavements may have erosional actions going on that are relatively unnoticed until physical cracking and deterioration show up. Figure 2.8*a* shows the pumping of fines out of a saturated AC

**FIG. 2.7** One-year-old interstate highway in western state; PCC on CTB; light stains indicate bleeding and pumping; new in 1970; May 1971.

pavement on an interstate highway in the Pacific Northwest. This section is on a fill, with no groundwater inflows. During the site inspection, close observation of the pavement under passing trucks showed on oozing of fines and water out of the cracks. This is definitely an example of the pumping of material out of the base, subbase, or subgrade of an AC pavement. Figure 2.8*b* shows excessive bleeding and staining on an airfield pavement in an eastern part of the United States. When conditions such as those shown in Fig. 2.8 can be seen, the serviceability of the pavements is being reduced by the combined actions of traffic and water.

When heavy wheel loads pass over saturated concrete pavements, it is often possible to see water and fine suspended matter being squirted out of cross joints and cracks. This behavior has been widely observed in highways for the past several decades; the author has talked with airport engineers who have seen "geysers" rise higher than the wings of large planes.

Figure 2.9*a* shows matter being squirted out of joints in a PCC pavement in a semiarid climate (rainfall about 12 in./year). This pavement was built in about 1950 and was several years old at the time

**FIG. 2.8** Losses of base materials by pumping and bleeding often shows up by staining. (*a*) Light stains at cracks in AC show loss of material by pumping; 16-yr-old interstate highway in Pacific Northwest; no groundwater; October 1970. (*b*) Light stains and water show bleeding of airfield runway; is from rainwater, as no groundwater exists here; an eastern state; about 20 years old; August 1972. (Courtesy CERL.[1])

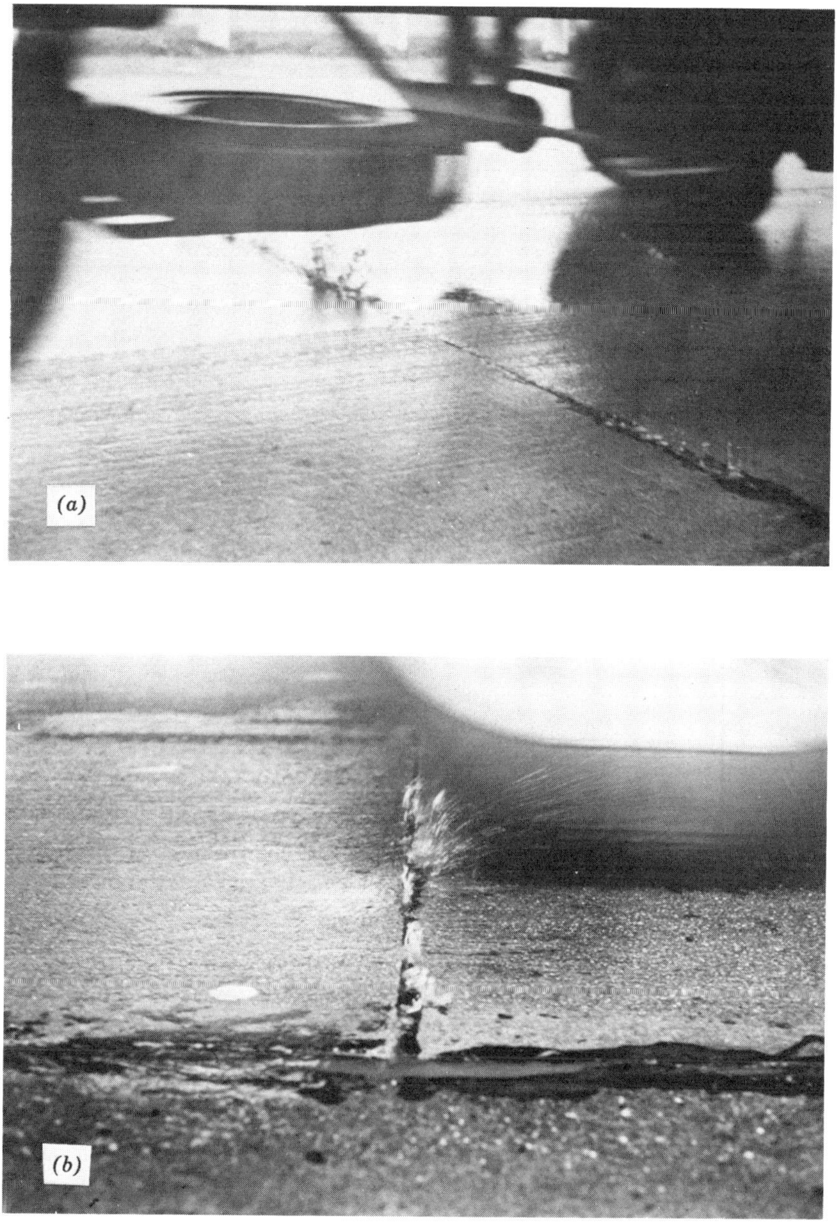

**FIG. 2.9** Stabilizing bases may not stop pumping. (*a*) Water and base material squirting up through joint in PCC pavement on untreated aggregate base, under impact of truck. (Courtesy, California Department of Transportation, photograph by C. W. Clawson, about 1960.) (*b*) Water and other matter squirting out of joint in "improved" design with PCC on CTB; much deterioration of this 12-year-old interstate highway pavement on 50-ft-high fill; no groundwater; photograph taken in 1970.

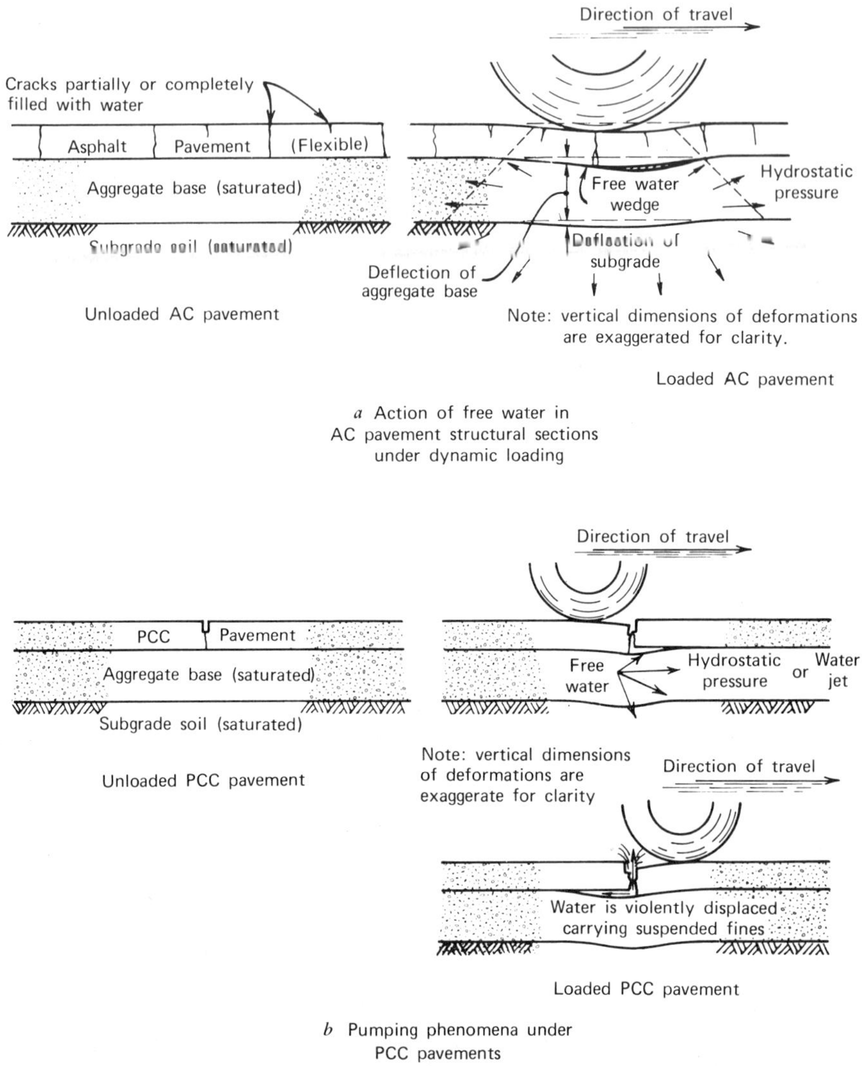


**FIG. 2.10** The ways that traffic impacts can damage saturated pavements (both rigid and flexible) (from "Final Report", FHWA, 1973b).

of the photograph. It was constructed with an 8-in. concrete slab on an untreated aggregate base. Figure 2.9*b* shows the squirting of matter out of a cross joint in an interstate highway pavement that was built in about 1958 on a cement-treated base to try to reduce water damage. Severe deterioration of significant lengths of this highway had taken place after less than 10 years of service. This "improved design" did not eliminate the primary cause of pavement deterioration—excess water.

Sketches in Fig. 2.10 illustrate some of the possible ways that heavy wheel impacts on saturated pavements can damage both AC pavements and PCC pavements. Vertical dimensions of cavities and deformations have been exaggerated for clarity.

Whenever the signs of bleeding, surging, and pumping, as illustrated here, are seen in pavements, it is evidence that these pavements are having to carry traffic at least some of the time while they contain excess water. While it is raining, it is often difficult to detect bleeding, pumping, and so on, because water is in such abundance everywhere. Shortly after a rain, however, it is often possible to see water "bleeding" by gravity out of the lower edges of pavements (Figs. 1.2*a* and *b*), or being pumped out by traffic (Figs. 2.9*a* and *b*). Often, while structural sections are still well filled with water (even several days or a week after a rain), by close observation it is possible to see the water level rise and fall in outside joints during the passage of trucks or other heavy vehicles. After the pavement surfaces and shoulders have partially or completely dried, stains left by dissolved or suspended matter are telltale evidence of pumping (Figs. 2.7 and 2.8).

## 2.4 EXCESS WATER AND SHORTENED PAVEMENT LIFE

During the times that pavements contain free water, one or more of the damaging actions described in Sec 2.2 take place at a much faster rate than when no free water is present. Some can occur *only* when there is free water to be acted on (pumping, erosion, frost action, etc.).

The net effect of the deteriorating actions of traffic and water and other elements usually can be assessed only by the apparent visible condition of most pavements, as expressed by a number or rating of its "condition" or "serviceability," at a given time. Among the factors that have been used in rating pavements are (1) amount of faulting, (2) number, width, and extent of cracks, (3) depth and extent of rutting, (4) other apparent damage, and (5) discomfort or hazard to the users of the pavements. If detailed records were kept of such significant factors for actual pavements in relation to environmental conditions and number of wheel loads, it would be possible to assess the relative rates of damage during both "good" and "bad" environmental conditions. Also, if maintenance man-hours and costs were recorded in relation to amounts of traffic during wet and dry conditions, assessments could be made of the relative rates of damage. These kinds of data would permit studies to be made comparing the performance and total cost of conventional slow-draining pavements with well-drained pavements.

Relatively little information of this kind is being obtained; however, a number of states and other agencies have regular programs for evaluating the condition of selected pavements several times a year. Housel (1962) points out that pavement condition surveys dating back to the middle twenties in Michigan have shown significant correlations between pavement condition and environmental factors, including soil type, drainage, and climatic factors. Extensive field measurements with the Michigan Truck-Mounted Profilometer have shown significantly greater increases in pavement roughness in wet periods than in dry. A number of other types of testing devices are used by other agencies and states to measure pavement condition at various times of the year, and a good deal of information is being gathered that should be of value in assessing the effects of environmental factors, such as excess water on rates of loss in pavement serviceability.

In the studies that were made for development of the *Guidelines* (FHWA, 1973a), the term "severity factor" is used in comparing damage during wet and dry periods. This factor is defined as the "relative rate of damage per load application" during a given environmental condition as compared with another. Thus if damage when free water is present is 5 times greater than when there is no free water, the severity factor is 5.0.

A detailed study of performance records, maintenance costs, and the like in relation to climatic conditions could provide realistic severity factors for actual pavements at given times in their life cycle. Seldom are such records kept or analyzed in this way; however, detailed records that were kept for a number of controlled road tests do provide a basis for such determinations (Table 2.1). For example, a special Highway Research Board report (HRB, 1955) states that during the period of worst damage to the WASHO road test in Idaho, the rate of distress averaged 748 $ft^2$/day for 12 load applications per day; in the period of least damage, the rate of distress was 1.0 $ft^2$/day for 1173 load applications per day. The *relative* rates of damage for this case can be calculated as $(748/12)/(1.0/1173) = 62/0.00085 = 70{,}000{:}1$, for a severity factor of 70,000 during the worst periods as compared with the best. In this comparison, the relative damage per application during the best period is 1.0.

**TABLE 2.1** Severity Factors in Experimental Road Tests As Estimated from Published Reports[a]

| Test | Reported by | Behavior reported | Severity factor |
|---|---|---|---|
| WASHO road test | HRB Special Report 22, 1955 | During worst period,[b] the rate of deterioration averaged 748 $ft^2$/day for 12 load applications per day; during best period, the rate of deterioration was 1.0 $ft^2$/day for 1173 load applications per day | 70,000:1 |
| AASHO road test | W. J. Liddle, p. 40, *Proceedings, First International Conference on Structural Design of Asphalt Pavements*, Ann Arbor, Michigan, 1962 | Deteriorating effects of traffic less severe in summer and fall than in spring. Road test data indicate a regional factor of 0.3 to 1.5 should be applied to loads on dry roadbeds (summer and fall) and 4.0 to 5.0 to loads on saturated roadbeds | Practical value between 10:1 and 40:1 |
| University of Illinois circular test track | Ernest J. Barenberg and Owen O. Thompson, *Highway Engineering* Series No. 36, University of Illinois, January 1970 | 700,000 load applications (3200-lb load on single tire) produced 0.2 to 0.5 in. rutting of unsaturated roadbed; after saturation, 12,000 additional loads destroyed the pavements (test set no. 1) | 200:1 |

[a] Courtesy CERL.[1]

[b] The frostmelt period.

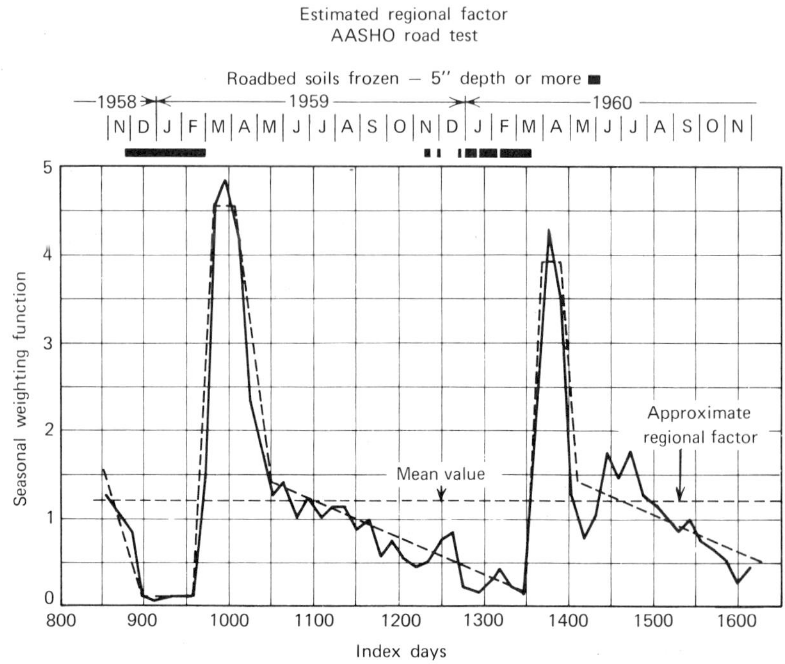


**FIG. 2.11** Seasonal variations in losses in serviceability of AASHO road test pavements as reflected by variable regional factor. Reprinted from the Proceedings of the First International Conference on the Structural Design of Asphalt Pavements (1962) by permission of W. J. Liddle and the Conference Executive Committee.

Liddle (1962) reports for the AASHO road test at Ottawa, Illinois that the damaging actions of traffic were noticeably more severe in spring than in summer or fall. Road test data suggest that a factor of 0.3 to 1.5 be applied to load applications on dry roadbeds (summer and fall) and 4.0 to 5.0 for loads applied to saturated roadbeds. He uses the term "regional factor" to express the number to be applied. In our discussion we are interested in the *relative* rates of damage under "wet" or super-saturated conditions as compared with "dry," which appear to be in the order of 5:1 to around 40:1 for the AASHO road test, depending on the individual values selected. Typical curves showing seasonal variations in losses in serviceability of pavements in the AASHO road test (as reflected by a variable regional factor) are given in Fig. 2.11.

Barenberg and Thompson (1970) report the results of accelerated traffic tests made in the University of Illinois Circular Test Track, which were intended to simulate conditions at the AASHO road test. Subgrade, base, and pavement layers were constructed in this test track using materials similar to those in selected test sections of the AASHO road test, the thicknesses and loads being scaled down to allow the tests to be completed in a reasonable length of time. In test set no. 1, a total of 700,000 load applications were applied (3200-lb wheel load on a single tire) to the test track at "as built" moisture content, and rutting of 0.2 to 0.5 in. developed in the traffic lane, with no evidence of pavement distress. The track was then exposed to surface sprinkling and subsurface soaking for 6 days, after which the saturating was stopped and the water level lowered to 1.0 ft below the top of the pavement. Traffic was then resumed, and 12,000 additional applications damaged the pavements to the extent that the testing had to be stopped. For this test, the rate of damage after soaking was about 200 times greater than before soaking; thus the indicated severity factor is 200. Figure 2.12 gives a typical curve showing the rate of loss of serviceability of one of the pavements in one of the test sets, as measured by the amount of rutting. From the start to point *B* the test track was at the "as built" moisture conditions; at point *B* water was added, as described; at point *C* the testing had to be stopped.

According to the records of these various controlled traffic tests, rates of damage for wet versus dry conditions ranged from a high of 70,000 to a low of perhaps 5 or 10. For actual highway and airfield pavements, the *severity factor* will vary with the components of the cross section, their thickness, the height of free water in the structural section, and the state of deterioration of the pavement. The cumulative

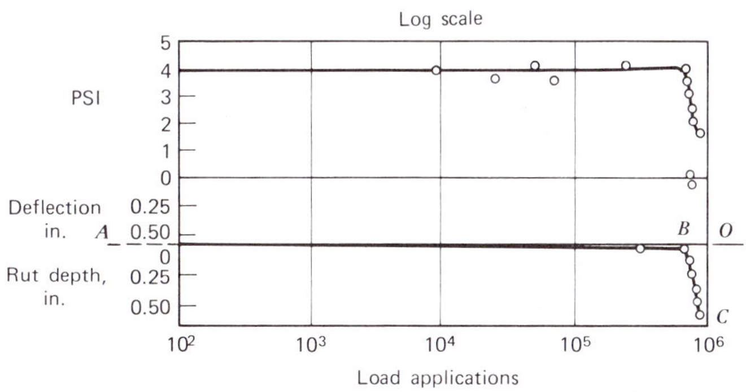


**FIG. 2.12** Typical performance data for University of Illinois circular test track (wet and unsaturated load applications compared), Barenberg and Thompson, 1970).

effects of load applications in periods during which the section contains excess water depend not only on the severity factor but also on the amount of time each year that the structural section contains free water. Realistic estimates of rates of deterioration of the pavements often can be made when pavements are requiring excessive maintenance, significant amounts are having to be dug out and replaced, or thick overlays are being required. By studying weather records the length of exposure to surface water can be added up. Then, by observing the behavior of the pavement (signs of bleeding at lower edges, presence of water in piezometers, etc.) the additional time-lag of drainage of excess water between rainfall events can be estimated. With this information estimates can be made of the rate of damage or amount of repair costs per load application in both wet and dry periods, and the severity factor of wet applications can be estimated.

To aid in appreciating the importance of the "severity factor" concept on pavement life, it is an easy matter to develop charts by mathematical computations that will enable design engineers to evaluate these effects on effective life cycles. Such a chart is given in Fig. 2.13 for a range of severity factors and amounts of exposure to free

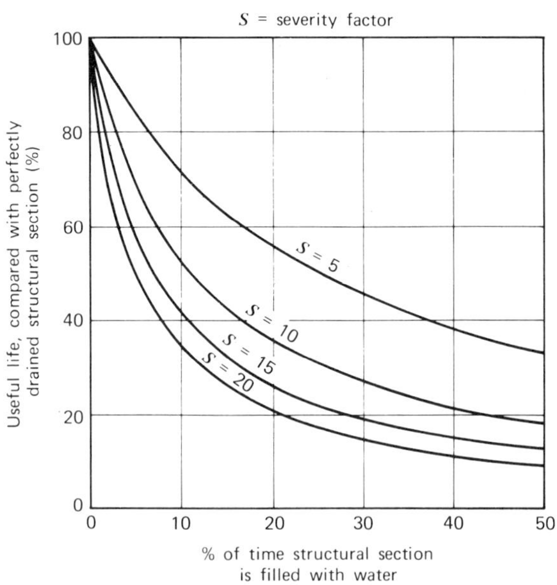


**FIG. 2.13** Chart showing effect of excess water on life cycles of pavements.

water each year. Points on this chart are determined by simple calculations. Thus if 80 percent of the impacts are "dry" with a rate of damage of unity and 20 percent are "wet" with a severity factor of 10, the equivalent number of dry applications would be (80)(1.0) + (20)(10) = 80 + 200 = 280, and the time in which this amount of loading would occur would be 100/280 = 0.35, or 35 percent of the time needed if all impacts were on a "dry" roadbed with a severity factor of 1.0. From Fig. 2.13 it is apparent that if pavements are required to carry heavy loads for even moderate amounts of time when filled with water, life cycles can be severely shortened. When environmental conditions vary from season to season each year, the rates of damage to pavements can be expected to follow cyclic trends, with greater amounts of damage and losses in serviceability occurring during wet periods than during dry. This, of course, is what actually happens. To illustrate the way that seasonal fluctuations might occur, Fig. 2.14 was prepared. Here three charts show the amount of damage that might occur to pavements in the wet and dry parts of a typical year. Thus the top chart (Fig. 2.14*a*) represents a pavement with a severity factor of 5 acting 2 months or one-sixth of the year, the middle chart (Fig. 2.14*b*) depicts one with a severity factor of 10 also acting one-sixth of the year, and the bottom chart (Fig. 2.14*c*) represents a pavement with a severity factor of 20 in effect one-sixth of the year, also. The percent of the annual damage in the wet period is shown on these charts. For simplicity of presentation, the wet impacts in Fig. 2.14 are all grouped together near the center, although they would normally be scattered out intermittently over the wet season.

From Fig. 2.14 it is seen that for $S = 5$, half of the annual damage would occur in the 2-month period of wet load applications; that for $S = 10$, the damage from wet applications during one-sixth of the year would be 67 percent of the yearly total; and for $S = 20$, the wet load applications would produce 80 percent of the yearly deterioration. In this drawing, the severity factor for dry loads (saturated subgrade, but no damaging free water in the structural section) is assumed to be 1.0.

The greater steepness of the curves in Fig. 2.14 in months 7 and 8 corresponds with the greater damage that normally occurs to many pavements during unfavorable environmental conditions, as already discussed. Some of the large-scale road tests had much greater damage in wet periods than is depicted in this figure. Recognizing that such variations do in reality occur in most pavements is an aid in understanding the role of water in pavement deterioration and the potential benefits of designing pavements as rapidly draining structural systems.

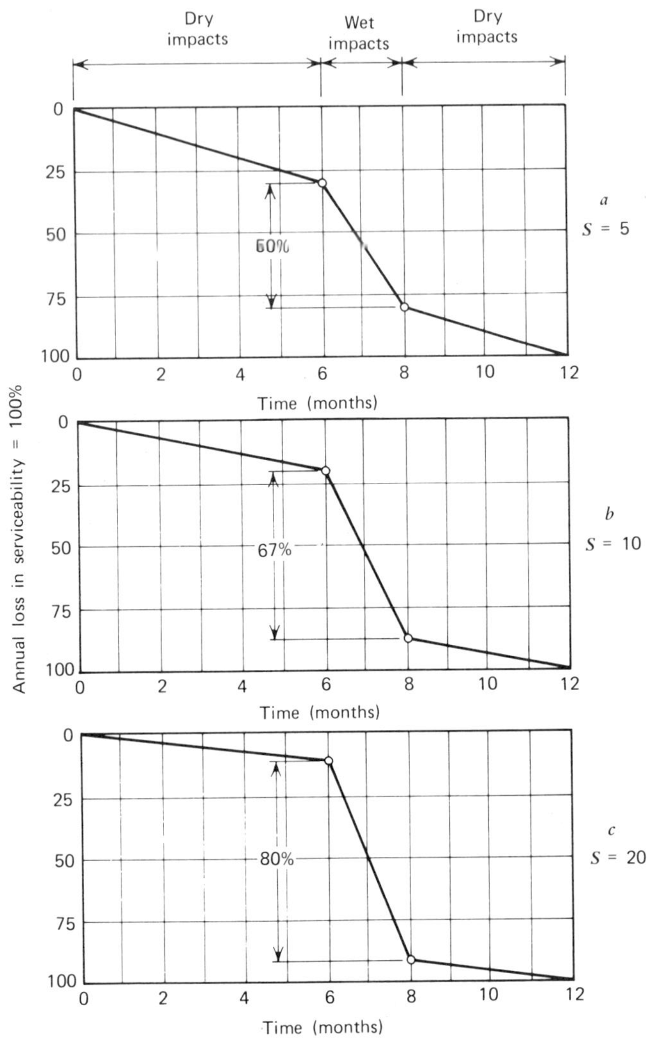


**FIG. 2.14** Chart showing losses in serviceability in wet and dry periods, for $S = 5$, 10, and 20.

## 2.5 THE SEVERITY FACTOR IN PAVEMENT DESIGN

Every pavement built has its individual structural strength capabilities and degree of resistance to all of the combined effects of environmental factors such as excess water, temperature, and traffic. At a given time in the life of a pavement, traffic impacts that occur when it contains excess water cause a specific rate of damage under the design wheel loads. Rates of damage will vary with the age and state of deterioration of a given pavement, dimensions and extent of cavities between pavements and bases, and the height of the free water level in its structural section. If a design formula allows fully for the variations that occur, the pavement should not become severely deteriorated or require excessive replacements, thick overlays, or major maintenance before reaching its design age. Inherently, any design method that properly allows for the variable environmental conditions that occur must include proper allowances for the accumulating damage that occurs over its full life cycle. Several examples are presented here to illustrate the way the serviceability of a pavement might decrease over its life cycle, assuming that the cyclic reductions in serviceability are allowed for in the design formula, and also assuming that they are not. Thus curve *A-A* in Fig. 2.15 shows a possible serviceability-time curve for a pavement with a design life and an actual effective life of 20 years. In developing this curve it was assumed that this pavement carries its traffic 10 months of the year in a dry condition and 2 months in a wet condition. Traffic is assumed to be increasing 5 percent a year over the 20-year period. For these assumptions, this pavement will

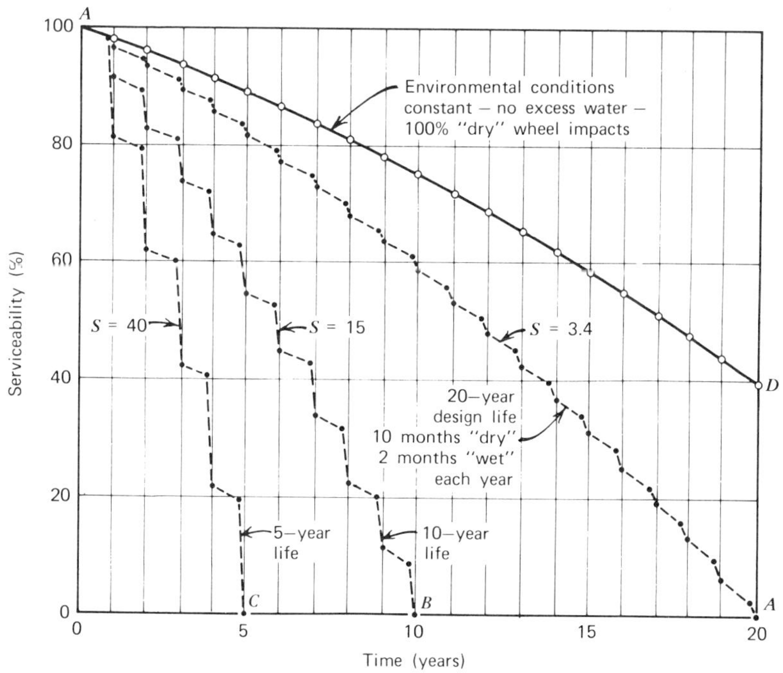


**FIG. 2.15** Diagram showing effect of excess water on pavement life.

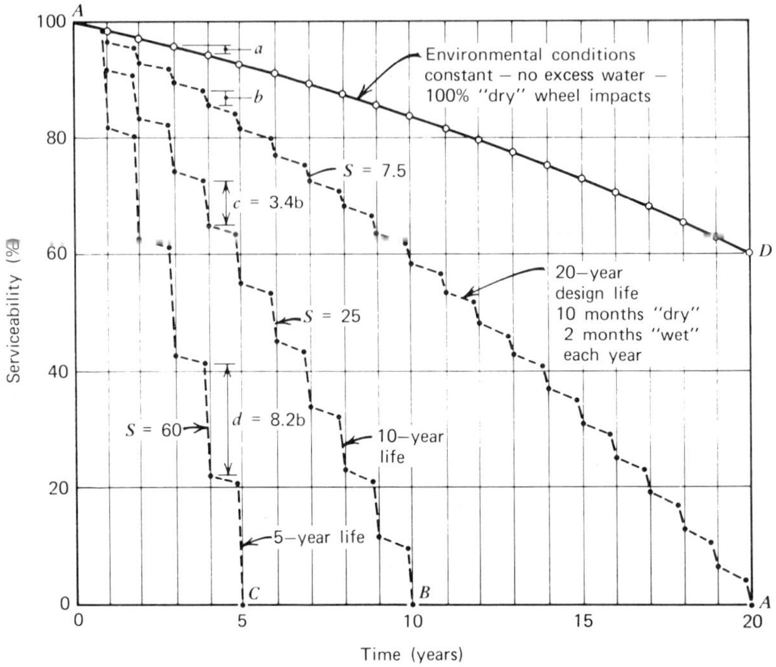


**FIG. 2.16** Diagram similar to Fig. 2.15, but for other conditions.

reach zero serviceability in 20 years if the severity factor, $S$, for wet impacts is 3.4. The formula has a factor of safety of 1.0 for one-sixth of the annual applications with a builtin severity factor of 3.4, and five-sixths of the applications with a severity factor of 1.0.

In developing Figs. 2.15, 2.16, and 2.17, the term "dry" is used to represent periods of load applications when there is no damaging free water in the structural sections, but *the subgrade is saturated.* The term "wet" is used for the periods of load applications when cavities, joints, spaces between pavements and bases, and pore spaces are essentially filled with water, providing free water to be acted on by dynamic wheel impacts.

Referring again to Fig. 2.15, suppose that this particular pavement never actually contains free water in significant amounts. In this case, the serviceability-time curve would be a smooth curve, without the steep segments in curve *A-A* during the wet periods, as there are no wet periods. It will follow line *A-D*. For this condition, 40 percent of its total serviceability will remain after 20 years, and it will have an actual effective life of about 28 years (extrapolating). But suppose that the true severity factor is greater than 3.4. If $S = 15$, this pavement will have an effective life of 10 years; if $S = 40$, its useful life will be only 5 years.

For comparative purposes, Fig. 2.16 represents a pavement with a theoretical and actual effective life of 20 years, with a builtin severity factor of 7.5 during one-sixth of the impacts per year. Now, if it were possible to keep all excess water out of the structural section, the serviceability-time curve would follow line *A-D*, and 60 percent of its serviceability would remain after 20 years. In this case its actual effective life would be in the order of 35 years. But if the actual "wet" damage is greater than allowed in the formula, the serviceability-time curve might follow line *A-B* (for an actual value of *S* of 25) or line *A-C* (for an actual value of *S* of 60).

Another illustrative example is given in Fig. 2.17. In this case the formula contains no builtin allowance for wet severity factors above 1.0, and the theoretical serviceability-time curve follows line *A-A*, with the serviceability dropping off to zero at 20 years. Now if the severity factor of wet loads taking place one-sixth of the time each year is 7, the

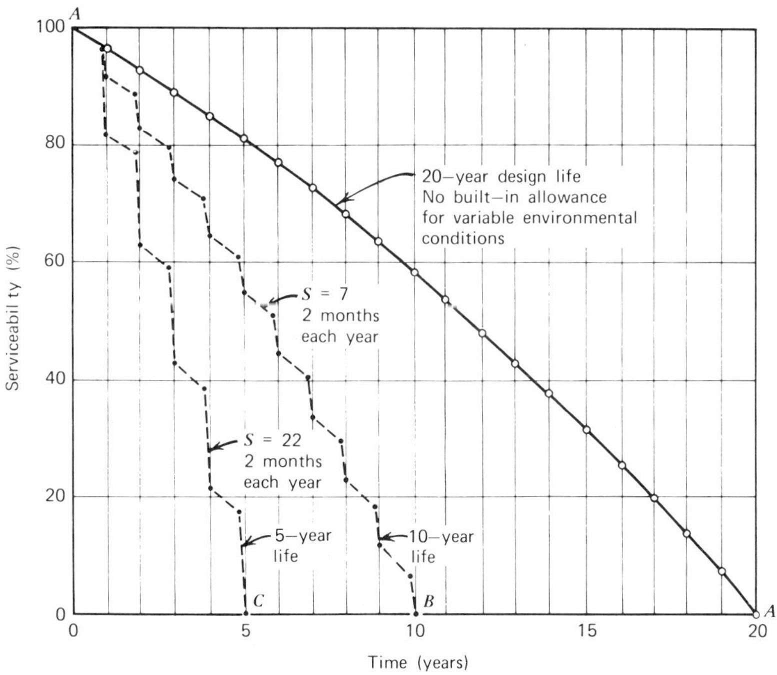


**FIG. 2.17** Diagram similar to Figs. 2.15 and 2.16, but for other conditions.

effective life would be reduced to 10 years; if the severity factor for wet loads is 22, the effective life would be 5 years.

In all these examples it is assumed that when a pavement's serviceability reaches zero, major restorations, overlays, or repairs will be needed to keep it in operation. As pointed out earlier, it is believed that rates of deterioration of most pavements (per load application) become progressively greater as they age under traffic and weather. In these examples, no allowance is made for increased rates of damage with age, and uniform rates are assumed throughout. Even with this assumption (which probably underrates the importance of wet impacts), substantial amounts of the annual losses in serviceability occur during wet periods of relatively short duration. For the assumption that these pavements contain excess water one-sixth of the time each year, the calculated percentages of the annual deterioration that would occur during this amount of unfavorable environmental condition would be as given in Table 2.2.

The information presented in Figs. 2.15, 2.16, and 2.17, together with Table 2.2, all point up the fact that if design formulas do not contain sufficient inherent builtin allowances for traffic damage during wet periods, pavement life can be greatly shortened, and excessive repairs, overlays, and replacements may be required before the design age is reached.

In consideration of the information just presented, it must be evident that if by any practical and economical means it would be

**TABLE 2.2** Amount of Annual Deterioration Occurring During Wet Wheel Impacts

| Example | Severity factor | Deterioration in wet periods (%) |
|---|---|---|
| Fig. 2.15, line *A-A* | 3.4 | 40 |
| Fig. 2.15, line *A-D* | 1.0 | none |
| Fig. 2.15, line *A-B* | 15 | 75 |
| Fig. 2.15, line *A-C* | 40 | 89 |
| Fig. 2.16, line *A-A* | 7.5 | 60 |
| Fig. 2.16, line *A-D* | 1.0 | none |
| Fig. 2.16, line *A-B* | 25 | 83 |
| Fig. 2.16, line *A-C* | 60 | 92 |
| Fig. 2.17, line *A-A* | 1.0 | none |
| Fig. 2.17, line *A-B* | 7.0 | 58 |
| Fig. 2.17, line *A-C* | 22 | 81 |

possible to greatly reduce or essentially eliminate the occurrences of excess water, pavement life could in many cases be greatly lengthened. For example, if all exposure of structural sections to excess water could be prevented in the case described in Fig. 2.15, the serviceability of the pavement would follow line *A-D*, and the life cycle could be extended to about 28 years (extrapolating). Likewise, complete elimination of excess water for the pavement represented by Fig. 2.16 would extend the life cycle to about 35 years.

This somewhat detailed analysis has been presented to show the major potential effects of a factor that is not normally considered in the design of pavements. It is presented to bring out the fact that any changes in design practices that would significantly reduce exposure of pavements to internal flooding offers the means for prolonging pavement life and saving literally billions of dollars in repair, overlay, and replacement costs of important pavements.

## References

Adams, L. G. (1969), "Report on Flexible Pavement Distress in the Atlanta Area," State Highway Department of Georgia, unpublished report, 1969.

Barenberg, Ernest J. and Owen O. Thompson (1970), "Behavior and Performance of Flexible Pavements Evaluated in the University of Illinois Pavement Test Track," Highway Engineering Series No. 36, Illinois Cooperative Highway Research Program Series No. 108, Urbana, Illinois, January 1970.

Federal Highway Administration, FHWA (1973a), *Guidelines for the Design of Subsurface Drainage Systems for Highway Structural Sections*, issued 1973 (also called *Guidelines*).

Federal Highway Administration, FHWA (1973b), *Final Report*, studies for the development of *Guidelines for the Design of Subsurface Drainage Systems for Highway Structural Sections*, Washington, D.C.

Finn, Fred (1962), Discussion, *Proceedings, International Conference on the Structural Design of Asphalt Pavements*, Ann Arbor, Michigan, August 20 24, 1962, p. 19.

Finney, E. A. (1973), "Better Concrete Pavement Serviceability," American Concrete Institute Monograph No. 7; issued jointly by the Iowa State University Press, Ames, Iowa, and the ACI, Detroit, Michigan, p. 69.

Frost, Harwood (1910), *The Art of Roadmaking*, The Scientific Press, Brooklyn, N.Y.

Highway Research Board, HRB (1952), "Special Report 4, Final Report on Road Test One-MD," Washington, D.C., 1952.

Highway Research Board, HRB (1955), "Special Report 22," 1955.

Highway Research Board, HRB (1962), "Summary Report (Report 7) the AASHO Road Test," National Advisory Committee, HRB, Special Report 616, Publication Number 1061, 1962.

Highway Research Board, HRB (1968), "State of the Art of Rigid Pavement Design," HRB Committee on Rigid Pavement Design, 1968.

Housel, William S. (1962), "The Michigan Pavement Performance Study for Design Control and Serviceability Rating," *Proceedings, International Conference on the*

*Structural Design of Asphalt Pavements*, Ann Arbor, Michigan, August 20–24, 1962, pp. 219–52.

Kirsten, M. S. (1944), "Survey of Subgrade Moisture Conditions," *Proceedings*, Highway Research Board, 24th Annual Meeting, Washington, D.C., p. 497.

Kirsten, M. S. (1946), "Subgrade Moisture Conditions Beneath Airport Pavements," *Proceedings*, Highway Research Board, 25th Annual Meeting, Washington, D.C., p. 450.

Liddle, W. J. (1962), "Application of AASHO Road Test Results to the Design of Flexible Pavement Structures," *Proceedings, International Conference on the Structural Design of Asphalt Pavements*, Ann Arbor, Michigan, August 20–24, 1962, pp. 42–51.

McAdam, John L (1820), "Report to the London Board of Agriculture."

McCullough, B. F. (1970), *Design Manual for Continuously Reinforced Concrete Pavement*, U.S. Steel Corp., Pittsburgh, Pennsylvania, 1970.

Myers, B., C. A. Hogentogler, E. S. Barber, and H. C. Porter (1941), "Discussion of Soil Water Phenomena," *Proceedings*, Highway Research Board, 21st Annual Meeting, Washington, D.C., p. 451.

O'Flaherty, C. A. (1967), *Highways*, Edward Arnold Publishers, London.

Porter, O. J. (1938), "The Preparation of Subgrades," *Proceedings*, 18th Annual Meeting, Highway Research Board, Part II, pp. 324–31.

Spellman, Don L. (1972), "Faulting of Concrete Pavements," *Highway Research Record*, No. 407, HRB.

Strassenmeyer, O. A., and J. O. Wilson (1967), "Some Observations of Edge Pumping," Connecticut State Highway Department Report (unpublished).

Verbeck, George, Paul Klieger, David Stark, and Wilmer Teske (1972), "Interim Report on D-Cracking of Concrete Pavements in Ohio," Agreement No. 1910, Ohio Department of Highways with Portland Cement Association, Research and Development Laboratories, Skokie, Illinois, March, 1972.

CHAPTER THREE

# Fundamental Aspects of Pavement Drainage

## 3.1 METHODS FOR CONTROLLING WATER IN PAVEMENTS

### Background

As with all other major classes of engineering structures having seepage problems, the damaging effects of water in pavements can be controlled if it is possible either (1) to keep the water out of places where it can cause damage or (2) rapidly and safely to remove by drainage methods that water which enters (Cedergren, 1967, 1974). An alternative to these two approaches is to do little or nothing about the water and to "beef up" the design to try to withstand the necessary loads in the presence of excess water.

In the field of dam engineering the conclusion was reached several decades ago that good drainage is essential to the safety and economy of earth embankment dams, concrete gravity dams, and other water-retaining structures. In the field of pavement engineering, however, the belief developed about 1940 that by using design formulas that make use of strength values obtained by testing saturated specimens of subgrade and base materials, any and all effects of water would be automatically accounted for and that rapid drainage of water out of pavement structural sections is not necessary. It was also felt that normal measures for sealing surfaces would keep enough of the surface water out of pavements and that drainage systems were needed only for control of high groundwater, springs, and other subsurface inflows. Nevertheless, there has been a continuing interest in drainage by a number of investigators down through the years, and numerous surveys and reports on prevailing practices have been published in the past 30 or 40 years.

Cotton (1937) presents a review of subdrainage practices, including the use of highway intercepting drains, airport surface drains, and highway subdrains to control free groundwater, longitudinal seepage, and so forth. Muir (1942) gives the results of a survey of the practices in all 48 states at that time. Keene and Horner (1951) present a detailed tabulation of replies to a Highway Research Board questionnaire about drainage practices. Nearly all the replies stated that faulty subsurface drainage caused serious types of pavement distress and failure, including rutting and shoving of flexible pavements, pumping in concrete pavements, and frost heave and boils in all kinds of pavements.

Tulayev and Prasad (1965) describe the methods used in the Soviet Union to stabilize roadbeds by drainage and to control the subsurface water regime. Drainage conditions in India are also described and analyzed in this report. The Indian Roads Congress (1967) presents discussions of road drainage practices in the United States, Canada, Czechoslovakia, Norway, the Phillipines, Malaya, the Netherlands, the Soviet Union, Spain, and the United Kingdom. Slaughter (1973) presents a very thorough summary of drainage methods in use in the United States and expresses the belief that there is a need for better utilization of problem-solving techniques for determining discharge requirements of subsurface drainage systems.

The Federal Highway Administration published its *Guidelines for the Design of Subsurface Drainage Systems for Highway Structural Sections* (FHWA, 1973a) after a thorough survey of modern pavements that are being damaged by surface water (FHWA, 1973, pp. 11–17 and 46–66). Detailed literature reviews for this study indicated that the prevailing intent in the design of modern highways has been to build them sufficiently strong to withstand the effects of water, but not to drain infiltered surface water. The comprehensive field inspections and interviews with knowledgeable highway engineers throughout the United States, together with nine case study site investigations, indicated that a great deal of the premature damage to modern pavements can be blamed on excess water that gets into structural sections—largely from surface waters. The *Guidelines* advocates designing subsurface drains as conveyors of water, using a designated percentage of the rainfall rate that can be expected frequently in a given climatic region and any other inflows from groundwater or other sources.

### Keeping Surface Water Out of Pavements

In the past several decades, a great deal of effort has been expended in trying to develop joint sealing systems and other surface sealing methods that would keep water out of pavements (Bruce, 1934; ACI, 1970).

Shirley (1970), in a discussion of flexible seals for joint movement in concrete structures, describes a wide variety of joint configurations and sealants in use or proposed. After listing the requirements for an ideal seal he comments, "It may one day prove possible to develop such a seal, but of those at present available, none possesses all the required properties to the necessary degree." Pole (1970), in a detailed review of a symposium on joints and sealants, presents calculated and measured movements at joints in PCC pavements. The actual movements (measured by E. Tons) show daily movements of about $\frac{1}{20}$ in. for a daily temperature variation of 30°F in 25- or 30-ft-long slabs. In regions in which the daily temperature variations are in the range of 50 to 60°, it appears that about $\frac{1}{10}$-in. movement could be expected—a very significant amount. Mr. Pole also comments, "There is clearly no perfect sealant and no ideal joint configuration."

Latter (1970) points out that variable friction on the underside of the slab and other nonuniformities can severely affect slab movement, with the result that one expansion joint may not move at all and the neighboring joint may take twice the planned movement, with consequent failure of the joint sealer. He says, "Despite many years of research and development of materials by sealant manufacturers and other bodies there is clearly no simple answer to the problems of joint sealing. By far the major problem is that of lack of adhesion over a period of time between the sealing material and the concrete face."

At the present time (1974) no practical and economical method has yet been developed that will keep pavements watertight for more than brief lengths of time after construction. Figure 3.1*a* shows an AC pavement in a hot, dry climate; the pavement has opened up largely because of shrinkage of the wearing course. Water that enters for brief periods each year has aggravated the structural problems in this pavement. Figure 3.1*b* shows a preformed rubberized joint filler that is being pulled out by traffic. Although the joint filler is still relatively "alive" after 3 or 4 years, it is no longer sticking to the pavement on both sides of the joint; this is one of the major problems that has not yet been solved—keeping joint fillers and sealants in contact with the pavement. Figure 3.1*c* shows a special preformed neoprene joint that has partially opened in a 4-year-old pavement in a southern state; this is allowing water to enter. Figure 3.1*d* illustrates the great difficulty of maintaining good joints between primary pavements and shoulders, an area in which serious problems often occur.

Nearly every pavement engineer has had the hope at one time or another that water will be kept out of pavements and that no special measures will be needed to drain out the amounts that enter. Figure 3.1

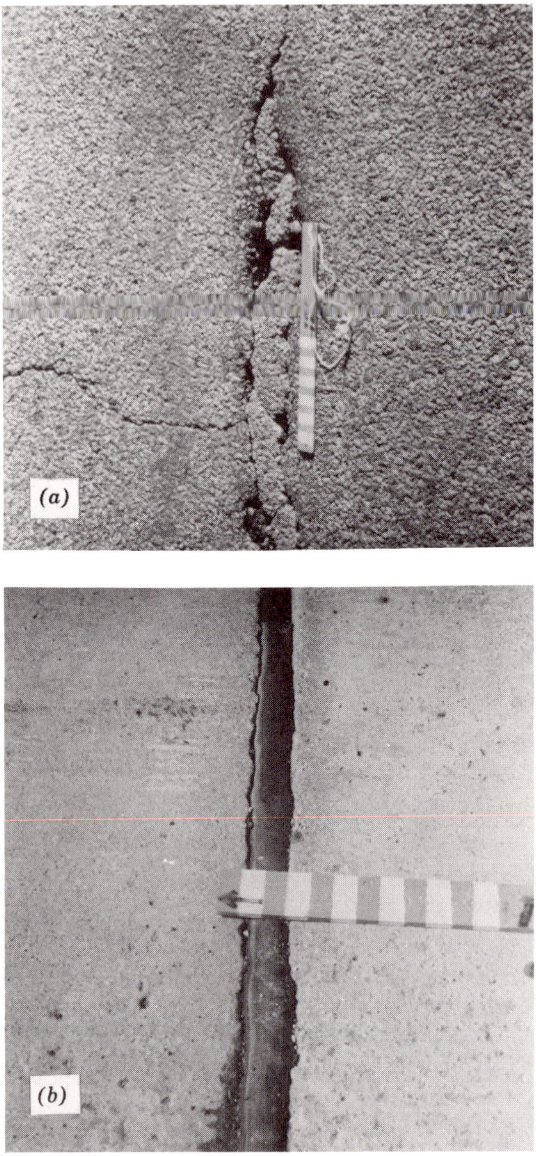


**FIG. 3.1** Keeping surface water out of pavements is difficult. (*a*) Cracks in AC pavement in hot, dry climate; infiltered water has added to problems; 1970. (*b*) Preformed rubber joint filler has been pulled out by traffic; 3 or 4-yr-old PCC pavement; Great Lakes area; October 1970. (*c*) Preformed neoprene joint filler has separated from concrete, allows water to enter; 4-year-old pavement in Southeast; Nov., 1970. (*d*) Deteriorated shoulder joint with pocket to trap water; high mountain area in western state; August 1970.

has shown some of the problems of keeping pavement cracks and joints watertight. Many relatively new pavements have, in fact, become quite permeable to the entry of surface water, as has already been brought out in Chapters 1 and 2. Unless good drainage systems are provided under pavements, serious damage under traffic usually occurs.

Cracks and joints of very small widths are capable of allowing surprisingly large inflows of water. In 1972 the author had the opportunity to assist with the performance of some inflow tests through some very modern joints in a runway-taxiway system constructed at one of the busiest commercial airfields in the United States. The runway-taxiway system had been dug out and completely reconstructed after the old runway had rapidly deteriorated. Those given the task of redesigning and reconstructing these pavements decided to develop an effective drainage system under the new facilities that would be capable of quickly removing any surface water that might enter through the top during the heavy rainfalls that occur frequently throughout the year. The 16-in.-thick PCC pavements they designed had construction joints every 25 ft in the transverse direction and every 75 ft in the longitudinal direction. Sawed joints were sealed with preformed polyethylene foam fillers and an FAA-approved poured filler.

The pavements were constructed on a drainage layer of ballast rock and crusher fines placed over the cement-treated silty clay subgrade. A drain system with collector pipes along the lower sides of the pavements fed the water to outlet pipes which discharged any water entering through the wearing course (or any other source) into manholes or on earth slopes off to the side of the paved areas.

After completion of the new runway-taxiway system, engineers noticed flows of water coming out of the drains during and after rainstorms. On further observations, they found that substantial flows began to emerge within the first hour after the start of a rainstorm, building up to a peak, and falling off to a trickle 10 to 20 hours after the end of the storm.

Even though one of the most modern types of joint sealing systems had been used, a substantial percentage of the rain that fell on these pavements entered apparently "sealed" joints and unsealed cracks, then flowed through the drainage system and out of the pipes. While visiting this airport in 1972 in a study of airport drainage for the U.S. Army Corps of Engineers (1974), the writer and an engineer of the local airport design office made some inflow tests of typical joints. In a typical test (see Fig. 3.2), a gallon jug of water entered 5 ft of joint in approximately 1 minute. This rate of flow represents an overall infil-

**FIG. 3.2** Making inflow test of joint in PCC airport taxiway pavement; heavy rainfall area; 1972.

tration rate (expressed as an average rate for slabs with longitudinal joints every 25 ft) of approximately 0.7 in./hour—which is a surprisingly large volume of water but which should probably not be unexpected in the light of research that was conducted by the University of Maryland with PCC pavements. A group of tests that is summarized in Table 3.1 illustrates the inflow rates that were obtained by the University of Maryland research. In these tests, PCC pavements with cracks of various widths were placed on a range of surface slopes. Precipitation was applied at a rate of 2 in./hour in each test, and measurements were made of the percentage of the runoff that entered the cracks.

The high rates of inflow given in Table 3.1 would apply to cracks that have no obstructions at the bottom. In many pavements, the rates of flow would probably be much less (for cracks of these widths) when the cracks and cavities under them became filled with water and back pressures or head built up.

Not only can water enter through cracks and joints, but relatively large amounts can percolate through new AC pavements that are constructed of some of the widely used "standard" types of mixes. Figure 3.3 shows a test of infiltration through a small slab of such a

**TABLE 3.1** Runoff into Surface Cracks[a] PCC Pavements (precipitation intensity—2 in./hour)

| Crack width (in.) | Pavement slope (%) | Percent of runoff entering crack |
|---|---|---|
| 0.035 | 1.25 | 70 |
| 0.035 | 2.50 | 76 |
| 0.035 | 2.75 | 79 |
| 0.050 | 2.50 | 89 |
| 0.050 | 3.75 | 87 |
| 0.125 | 2.50 | 97 |
| 0.125 | 3.75 | 95 |

[a] Research by University of Maryland (laboratory tests).

mix. This slab, approximately 8 × 10 in. in area, was placed hot within the frame and was compacted to about the density expected in the field. The amount of water flowing through this slab (as seen by the flow entering from the pipe, and dripping out below) represents a coefficient of permeability of approximately 100 ft/day (0.03 cm/sec), or 50 in./hour.

Numerous investigators have obtained coefficients ranging from a high of several hundred feet a day for unsealed AC mixes, down to virtually zero for well-sealed pavements. Obviously, the amount that can enter a given pavement depends on the kind of mix used, its density, degree of cracking, the kind and effectiveness of any sealers or joint fillers used, and the age and condition of the entire system. Surface openings and imperfections can make both AC and PCC pavements considerably more permeable than the bases on which they are constructed.

Most of the agencies responsible for maintaining important pavements have ongoing seal-coating and joint-sealing programs in an effort to keep pavements as well sealed as possible. However, some of those engaged in these programs say that they are involved in a difficult situation, and they flatly admit they have "given up" and are no longer trying to keep surfaces and joints sealed. Even though slurry seals, regular seal coats, and joint sealers become less effective after varying amounts of time (because of expansion and contraction, oxidation, traffic, etc.), it is still the considered judgment of the author that surfaces and joints of all important pavements should be kept as well sealed as is practical.

Several photographs in Chapters 1 and 2 show evidence of water getting into pavements, filling pore spaces, joints, and so forth, then bleeding or being pumped out of edge joints and transverse cracks and joints. Some additional pictures of such conditions are given here. Figure 3.4*a* shows a portion of an interstate highway in a semiarid climate in a western state, situated where only rainfall can enter this pavement. Here water is still bleeding and being pumped out by traffic 8 days after the last rain recorded by the U.S. Weather Bureau. This is a PCC pavement with a cement-treated base extending 1 ft beneath the shoulder. Although less than 2 years old at the time of the photograph, ⅛ in. of faulting has already occurred.

Figure 3.4*b* shows another pavement of similar design, also several days after a rain. This road is also on a shallow fill. The stains indicate losses of material by leaching or physical erosion and ejection with the water that escapes across the shoulder. Figure 3.5 shows the saturation level in an observation well in an interstate highway pavement at a time when internal saturation is comparable to that of Fig. 3.4*b*. The

**FIG. 3.3** Making percolation test on small slab of a "standard" AC mix; permeability about 100 ft/day; July 1973.

**FIG. 3.4** Without good drainage, flow of water out of structural sections can be very slow. (*a*) Heavy bleeding 8 days after a rain; PCC pavement on CTB; completed in 1970; ⅛-in. faulting; May 1971. (*b*) Strains across shoulder show loss of material by pumping and bleeding; new PCC on CTB; a few days after a rain; May 1971.

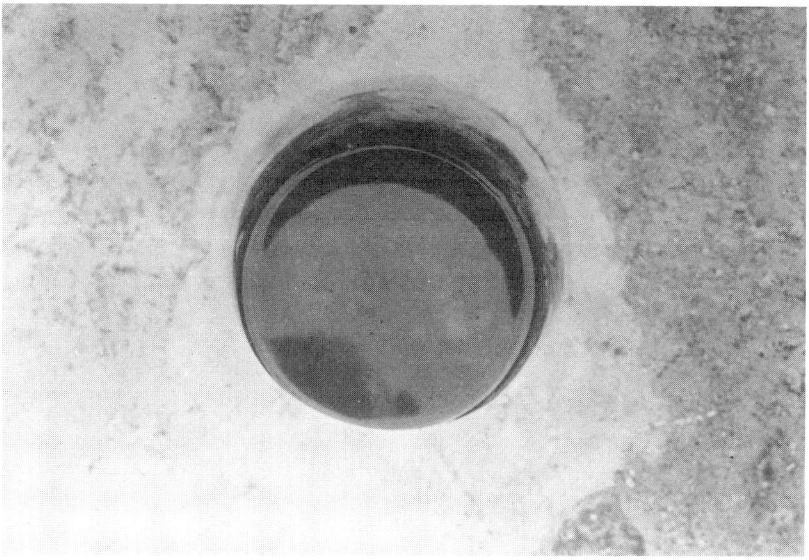

**FIG. 3.5** Saturation level in test hole in interstate pavement about a week after last rain; is comparable to condition in Fig. 3.4*b*. December 1970 (Courtesy, FHWA).

"artesian" head on the bottom of the pavement caused water to rise to within 2 in. of the top of the pavement.

Figure 3.6*a* shows a condition that is a common site at many airfields throughout wet climatic areas. Here, a healthy growth of grass is in a transverse joint, giving evidence of a good water supply and indicating that the structural section is in the "bathtub" condition already noted. Figure 3.6*b* shows cattails that are kept in a perennially damp condition by seepage from a "daylighted" sandy subbase of an interstate highway in the Great Lakes area.

All the photographs in this group show evidence of substantial supplies of water within pavement structural sections, which are furnished by inflows through their top surfaces. If the tops could be kept watertight, most of these occurrences of water could be prevented. The conditions shown in Fig. 3.6*a*, however, could be aggravated by flow of water into this pavement from saturated earth at its outer edge. This situation represents another common source of water in pavements.

## The Need for Drainage

If it is recognized that water cannot be kept out of pavements and if it is admitted that this can greatly shorten pavement life, it should be

**FIG. 3.6** Grass and weeds are other signs of plentiful water. (*a*) Healthy grass growth at edge of airfield taxiway and in joints shows liberal water supply and indicates poor drainage; Great Lakes state; September 1972. (Courtesy CERL.[1]) (*b*) Cattails growing on interstate highway slope; watered by seepage from "daylighted" subbase; Great Lakes area; October 1970.

concluded that pavement design procedures should produce systems ensuring that excess water will be rapidly drained out of structural sections.

Subsequent chapters of this book give detailed procedures for analyzing the drainage requirements of highway and airfield pavements and for designing the most optimum kinds of systems. When these systems are properly designed, constructed, and maintained, they can prevent prolonged accumulations of free water in structural sections and can virtually preclude damage from excess water, thereby greatly extending the useful life of most pavements. Even the most minimal system designed with the help of the rational methods described will have drainage capabilities hundreds of times greater than those of most of the pavements that have been built in the past 30 or 40 years. Consequently, when these methods are used, even if rough estimates have to be made of inflows, the resulting systems will usually be at least two orders of magnitude better than those designed without an analysis of discharge needs.

## 3.2 NEED FOR RATIONAL METHODS IN DESIGN OF DRAINAGE SYSTEMS

In the past subsurface drainage of pavements has generally been looked on as a qualitative problem with general solutions, when in reality every seepage and drainage situation has a specific, numerical solution (Cedergren, 1960; Lovering, 1960; Cedergren, 1967). Every body of soil or other porous matter has exact, specific water-transmitting capabilities at a given time for a specific set of conditions. Thus when a farmer has a well drilled, the well penetrates a complicated but individual layering of porous and nonporous formations, which under a given set of conditions allows his well to produce a specific amount of water. As long as nothing changes the characteristics of the water-bearing formations, the height of the surrounding water table, or the characteristics of the well, this well can be counted on to produce the same, identical quantity of water year after year. Similarly, every pavement has its individual characteristics at a given time and can accept certain, definite amounts of inflows through its networks of cracks joints, porous areas, and so forth. Likewise, its base drainage layers (if it has any) and its subgrade are capable of draining away certain quantities of water as long as the basic factors affecting the flow are not changed. In other words, every hydrodynamic flow situation is exactly *repeatable* if conditions are kept the same.

Even though the factors entering into seepage phenomena may not always be known with a high degree of accuracy, it is still worth the effort to make the best evaluation possible, rather than to depend on intuition or simply follow past practice in designing systems for engineering works. In the first place, pavements are inherently difficult to drain, and in the second place, intuition, rules of thumb, or guesswork can give highly undependable results, as is brought out in this chapter.

### Inflows Often Exceed Outflow Capabilities

Pavements for highways, airfields, large parking lots, city streets, and the like have large areas exposed to the weathering actions of nature, large areas exposed to water infiltration from a number of sources, and only small areas through which water can drain out. Thus the *geometrics* of wide, modern pavements compound the difficulties of draining these structures. Minor (1961) emphasizes that wide, multilane freeways are proving to be much more difficult to drain than older, narrower highways in the state of Washington.

Because of their wide, flat dimensions, pavements are highly vulnerable to inflows from many sources, such as (1) downward flow through porous surfaces, open construction joints, open shrinkage cracks, and so on, (2) lateral flow into the sections from water ponded on high medians, flooded outer edges, or unsealed shoulders, (3) upward flow from high groundwater, springs, river seepage, and so forth, (4) capillary suction from underlying water tables, and (5) condensation or "water of hydrogenesis" as a result of fluctuations in atmospheric temperature and pressures. Even localized inflows from some of these sources can cause accumulations of water in structural sections that can lead to failures having consequences far out of proportion to the amount of area affected.

Essentially, all pavements are exposed to surface infiltration from precipitation over entire expanses of wide paved areas. Since large hydraulic gradients are usually working to cause flows into pavements or bases, the quantities of inflow tend to be large.

In contrast, the thin vertical dimensions of pavement bases and drainage layers greatly limit the amount of area through which water can drain out. Likewise, the discharge hydraulic gradients in bases and drainage layers cannot exceed more than 0.01 or 0.02 in many cases; hence the combined effects of small area and small gradient greatly restrict drainage out of a structural section. The result is that the *outflow capability* is often much less than the *inflow rate,* which leads to the "bathtub" condition discussed in Chapters 1 and 2.

In addition to the natural sources of precipitation and surface inflow, some pavements are subjected to man-made supplies that can keep structural sections filled with water throughout the "dry" parts of the year. Water used for washing aircraft can keep service aprons wet a great deal of the time. Also, in dry climates sprinkling of freeway landscaping often results in thorough wetting of adjacent pavements and produces conditions in them similar to those created by natural heavy rainfall at frequent intervals. Figure 3.7 shows a state highway that is being thoroughly inundated *in the dry season* by the irrigation of a wide, grassed median.

When pavements are constructed in hilly terrain or in cut sections below the water table, spring inflows and seepage from groundwater can add to the supply entering these pavements. Figures 4.12 and 4.18 show water from springs bleeding out of pavements in hilly terrain. Figure 3.8*a* shows profuse flows onto a city street and adjacent sidewalks and properties caused by seepage from a nearby river at a high flood stage. This condition develops only when the river is above its normal levels. During these periods (which occur nearly every year) the pavements are virtually in a condition of flotation from the excessively high groundwater level. If no drainage is provided, as at

**FIG. 3.7** State road in West Coast city being watered by sprinklers used for irrigating wide median; dry season; 1971.

**FIG. 3.8** Simple drainage system relieves severe seepage condition. (*a*) Undrained city street and adjacent walks and land being saturated by seepage from adjacent river; spring, 1970. (*b*) Another area, closer to river, but with $2\frac{1}{2}$- to 3-ft deep drains along edges of street and between property lines; shows remarkable control at fairly high river stage; same day as photograph in Fig. 3.8*a*.

this location, this extreme seepage condition exists. Without drainage, water has access to the entire underside of the pavements wherever these artesian conditions exist. Severe though this condition appears, it is sometimes very easy to correct with drainage. Figure 3.8*b* shows a similar area (even closer to the levee) where trench drains $2\frac{1}{2}$ to 3 ft deep were installed along both sides of the street and between adjacent property lines. This comparatively nominal and inexpensive drainage system, with pipes discharging into a storm sewer, keeps the street and the adjacent walks and properties completely dry. These two photographs (taken on the same day) are a vivid illustration of the benefits of subsurface drainage.

Another more common source of subsurface water, that supplied by melting ice during the spring thaw, is illustrated in Fig. 3.9. In nearly every cold region of the United States, excess water rather suddenly becomes available in the spring, and if outflow capabilities of the subgrade and lateral drainage of bases are inadequate, pavements and bases can be in a flooded state for significant periods of time every year. This kind of bleeding, which looks just like that from rainfall, spring inflows, and the like, is an indication that outflow capabilities are not sufficient to keep up with the supply.

**FIG. 3.9** Excess water bleeding out of pavement in spring thaw; an eastern state (courtesy of George M. Ring).

Usually when pavements are built below the groundwater level or spring inflows are anticipated in hilly terrain, subsurface drains are provided. Such groundwater conditions prevail in only a small percentage of the total pavement systems; however, a far more common source of water is rainfall, which occurs in significant amounts on perhaps 90 to 95 percent of the important pavements in the United States.

Unless good drainage is provided, rainfall can completely fill all of the joints, cracks, pore spaces, and so forth in a comparatively short time. The author has witnessed "springs" coming out of the lower edges of wide airport pavements after light showers that produced only 0.02 to 0.03 in. of rain following a dry spell of a week. Figures 3.4*b*, 2.7, and 8.5*c* show stains that remain after such "spring" areas have completely or partially dried up. Phenomena such as these are in part caused by the geometric conditions that allow large amounts of water to enter pavements while greatly restricting the outflow capabilities.

Barber (1962) points out that special subsurface drainage was required at Idlewild Airport for removal of surface water infiltration, even though the natural subgrade was clean, relatively permeable sand. Because of large dimensions of the areas infiltered, the natural transmissibility was less than rainfall. For the wide dimensions of these areas, a clean sand with a coefficient of permeability of 100 ft/day would be capable of removing only 1.8 in./year on these wide areas at this airfield.

Adding to the inherently difficult problems caused by poor geometrics is the practice of building pavements in trenches, which often restricts the primary drainage to that which escapes through the top (see Figs. 1.2, 2.7, and 3.4*a*). Vallerga (1959), stresses the importance of providing drainage for the various layers of highway pavement structural sections in addition to the normal surface and foundation drainage systems. He warns of the trench ("bathtub") condition developing in structural sections enveloped beneath and laterally by materials that are not free draining when adequate outlets are not provided. He says that any design feature which would trap water at any point in the structural section should be avoided.

Izzard (1944) also stresses that drainage is essential when a pervious subbase is constructed in a trench section and water has access to the subbase through joints or cracks in the pavement or along the edge of the pavement. He describes a method for computation of the time for drainage in terms of permeability, slope, and the amount of drainable water.

Porter (1938), in discussing severe losses of support of pavements in Texas in the 1930s states, "Proper drainage has been considered the major problem in railway and highway construction and maintenance, but for a long time it was thought of as surface or external drainage. In order to maintain the soil structure at as nearly a uniform moisture content and volume as practicable, internal drainage must be given proper attention . . . ." Also, "The placing of coarse-grained material on clay soil and the flanking of the coarse grained material with tight clay soil probably has caused more pavement failures than any other one item . . . ." This kind of entrapment of water often occurs where roads are flanked by the usual "standard" types of low permeability granular materials used in shoulder construction, where such roads are on relatively impermeable subgrades. When these conditions exist, the water that enters can maintain the "bathtub" condition for weeks and months after every heavy rainfall event or series of events.

In the past efforts have frequently been made to drain pavements *after problems have developed* by the use of trench drains along outer edges of pavements. Narrow drains of this kind are usually relatively ineffective in improving roadbeds because their zone of influence is so small (see Fig. 1.8). Lancaster and Davis (1956), in their report on studies made in Missouri to determine the effectiveness of underdrains *installed during construction* of PCC pavements, show that these widely spaced, narrow, transverse trench drains were very ineffective in draining the sandy gravel base courses. The trench drains, about 18 in. deep and 18 in. wide, contained pipes that were surrounded with concrete sand backfill. Only minuscule quantities of water came out of the drains. When the condition of the pavements having these drains was compared with the pavements having no drains, it was not possible to detect any difference between the undrained and "drained" sections. This study demonstrates conclusively the *ineffectiveness* of narrow drains constructed with low-permeability drainage materials and emphasizes the importance of analyzing the outflow needs of drains.

### Rules of Thumb and Intuition Can Mislead Designers

Trying to judge the permeability requirements of drainage materials on the basis of general experience, intuition, past practices, or rules of thumb without the use of the rational approach can be very misleading. McClelland and Gregg (1944) decry the dependence of highway engineers on rule-of-thumb procedures for subsurface drainage design, and they advocate the use of more direct and improved analytical methods, such as flow nets, for analyzing drainage needs. They conclude that the use of the analytical methods would make subsurface drainage installations *more effective* and *more economical*.

The basic fallacy of relying on intuition, judgment, and so on when dealing with seepage and drainage can probably be blamed in a large measure on the extremely wide variability of the property permeability itself. There is no other engineering property of construction materials that has so wide a range. Coarse macadam aggregate can have coefficients of permeability in the order of 100,000 ft/day (30 cm/sec) or more, whereas fat clays often have coefficients in the order of 0.000003 ft/day ($1 \times 10^{-9}$ cm/sec), for an overall variation of several billion times. It is almost beyond human capabilities to appreciate or have a "feel" for the physical significance of anything having such an extreme range. When the further variable of the hydraulic gradient causing flow is included, the potential range of seepage velocities is in the order of 100 billion to 1 trillion times.

Obviously, when dealing with phenomena that can have so wide a range as seepage, relying on "intuition" can be highly misleading.

A vivid illustration of the possible wide range in the permeability of roadbed materials can be obtained by watching field permeability tests being made in holes drilled into the bases of typical pavements. Figure 3.10*a* shows such a test underway on a conventional, well-graded aggregate base of an 8-year-old pavement showing distress. After 100 minutes the water level in the test hole had gone down about ⅛ in. This is fairly typical of a number of such tests that have been made by the author in "conventional" highway and airfield pavements in widely separated areas of the United States. In sharp contrast is the test shown in Fig. 3.10*b*, in which a 2-in.-diameter hose was needed to maintain a head of a few inches on the very coarse macadam base under the Humboldt County "experimental" pavement described by Cedergren and Lovering (1968). Here the "special" base drainage layer has a coefficient of permeability of about 30,000 ft/day. The test depicted in Fig. 3.10*a* could have been kept supplied with an eyedropper, the second needed a flow of 75 gpm. The vast difference between the "conventional" base and the highly permeable base is very obvious; yet this represents a difference in the order of 1,000,000 times, a comparatively narrow part of the overall range that is possible.

In the past selection of drainage aggregates on the basis of general experience has led to the widespread usage of blends of sand and gravel with permeabilities in the range of 5 to 10 ft/day for pavement drainage. The very limited capabilities of some of these materials for draining pavements is illustrated by Table 3.2. In this table the

**FIG. 3.10** Field percolation tests in pavements show up wide range in drainability. (*a*) Gravel base under a PCC pavement in New England deteriorating from excess water; after 100 minutes water level has gone down ⅛ in.; very slow drainage; October 1970. (*b*) Open-graded macadam base under Humboldt County experimental pavement; in-place permeability is 30,000 ft/day for 75 gpm inflow; April 1971.

**TABLE 3.2** An Illustration of the Small Drainage Capabilities of Some Standard Base Aggregates[a,b]

| Class of material | Perme-ability (ft/day) | Rate of infiltration base can remove from structural section (in./hour) |
|---|---|---|
| Well-graded blend of sand and gravel | 1 | 0.0006 |
| Concrete sand with high dust content | 2 | 0.0012 |
| Concrete sand with low dust content | 20 | 0.012 |

[a] Four-lane road; discharge capabilities of bases estimated with Darcy's law, $Q = kiA$; thickness of base = 12 in.
[b] From H. R. Cedergren, *Seepage, Drainage, and Flow Nets*, John Wiley & Sons, Inc., New York, 1967, p. 371.

amount of water that can drain to outside edges is given in terms of the rate of infiltration the bases can remove from structural sections. It is for four-lane roads sloping to each side from centerline.

Blends of sand and gravel with the permeability characteristics of those in Table 3.2 are often used for bases and subbases in the hope that they will serve as "permeable" materials to protect the roads from damage from excess water. When no calculations are made of potential inflows or outflow capabilities, the tendency has been to underestimate the amount of water that will get into pavements and to overestimate the outflow capabilities. How, otherwise, can it be that pavement surface courses with effective coefficients of permeability in the range of 25 to 250 ft/day are regularly placed over bases and subbases with permeabilities in the range of 0.01 to 10 ft/day? In reality, if outflow capabilities are to be in balance with inflows, the base drainage layers should have coefficients of permeability at least several hundred times greater than the wearing courses through which water can enter. Even relatively crude calculations of inflows with Darcy's law using estimated coefficients of permeability and hydraulic gradients can often point up gross errors in judgment about the capability and suitability of various classes of materials to drain pavements. Table 3.2 points up the error of assuming that base materials with coefficients of permeability of a few feet a day can drain roadbeds.

The fallacy of assuming that ordinary, well-graded blends of sand and gravel used for bases and subbases will provide beneficial drainage is also pointed up by Fig. 3.11. Here the rates of infiltration that can be removed by 12-in.-thick bases under four-lane roads are plotted as ordinates, and coefficients of permeability are abscissae. It is assumed that the bases drain to both sides, that the hydraulic gradient in the bases is 0.03, and that the total surface accepting infiltration is 34 ft wide on each side of the centerline. Darcy's law ($Q = kiA$) is used in estimating the discharge capabilities of the bases. In this figure some "standard" bases are compared with open-graded bases. It is seen that the normal types of bases and subbases have a *maximum capability* of 0.01 in./hour, which is a light drizzle, down to about 1/1000th of this amount. In comparison, the *minimum capability* of the open-graded bases is 1.0 in./hour (for $k$ = 3000 ft/day).

Lovering (1960) suggests that efforts be made to "evaluate the variables" affecting drainage, even though this may be difficult;

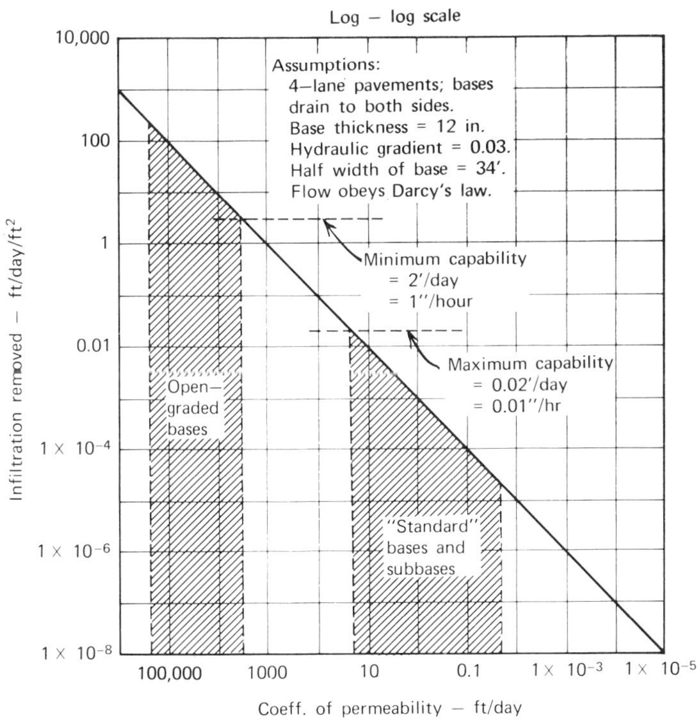


**FIG. 3.11** Infiltration-removing capabilities of some standard bases and subbases compared with open-graded macadam bases.

otherwise, inadequate drainage will result. He points out,

> If we assume a one-foot layer of filter material placed on a sixty-foot width of water-bearing soil with a 2% cross slope toward an underdrain (hydraulic gradient of 0.02), and if we assume, further, a coefficient of permeability of ten feet per day for the filter material, each lineal foot of this installation will be capable of carrying approximately 0.37 cubic feet, or less than 3 gallons of water to the underdrain each day.

He goes on to say, "A dripping faucet will produce a pint of water in 10 minutes, or 18 gallons in a day—six times as much water as would be removed by each lineal foot of underdrain conforming to the above conditions."

In sharp contrast with the optimistic idea of many engineers that concrete sand and comparable blends of aggregate can be used successfully for roadbed and airfield pavement drainage, agricultural experts look on *beach sands*, which are usually many times more permeable than concrete sand, as unsuitable for some agricultural uses because they *are too fine to permit free drainage* and thorough aeration.

Trying to use blends of sand and gravel for roadbed drainage is also incompatible with experiences in other major areas of civil engineering construction. Butterfield (1964) tells about the specifications the U.S. Army Corps of Engineers established for *the most impermeable zone* in the Howard A. Hanson Dam on the Green River in the state of Washington. The sand-and-gravel core of this dam was required to have a minimum fines content (passing the no. 200 screen) of 3 percent, which guaranteed a coefficient of permeability of not over $1 \times 10^{-3}$ ft/minute or 1.4 ft/day. Here, *a class of material approximately identical to one often used for draining pavements provides an effective water barrier which allows only limited seepage through a major dam*! The permeability needs or limitations for the material used in the Howard A. Hanson Dam were not left to luck or intuition; they were based on calculations of seepage rates possible with the materials specified.

### Direction of Drainage a Major Factor

When comparing the potential effectiveness of the various kinds of drainage systems that can be used for pavements it is important that the *direction* of the flow of water out of the primary pavement and base layers be considered, for this is a factor of major importance. If the way water can escape from structural sections is almost entirely through cracks and joints in the top of the pavement, as is the case in many of the photographs in Chapters 1 and 2, and in Fig. 3.4*a* and *b*, the pavements can often remain filled with water for weeks or months at a time. This is the most inefficient kind of drainage. Water cannot flow uphill under the forces of gravity, and it tends to migrate to lower elevations and outer edges, to sag vertical curves, and so forth, where it can often be seen "bleeding" out of cracks or joints or being "pumped" out under the pressures and impacts of heavy wheel loads. Sometimes thermal changes provide a boost in the form of an increase in the pressure in the air trapped under pavements, giving the water a "lift." (Large amounts of bleeding are often observed on airfield pavements during the warmer parts of the day, as seen in Fig. 3.12.) With all these helps, the structural sections can still remain saturated and contain excess water for prolonged periods of time after each rainfall, since drainage out of the tops of pavements is in opposition to gravity.

Where pavements on low-permeability subgrades are provided with drains along lower, outer edges of their bases, water that enters the structural section flows in a horizontal direction by *lateral drainage* toward the drains. This is also a relatively inefficient form of drainage *unless materials have very high permeabilities*, as can be demonstrated by Darcy's law. Discharge capabilities of bases are proportional to the lateral hydraulic gradient $i$, which is small—generally limited to the cross slope—and the thickness $A$, which is also a relatively small di-

**FIG. 3.12** Heavy bleeding of an airfield pavement (courtesy Waterways Experiment Station).

mension. Where the normal, well-graded, sandy gravel bases or subbases are used, their low permeability $k$ places another major restriction on flow possibilities. Nevertheless, even many of these materials would provide much better drainage if the water could flow vertically downward into a highly permeable base drainage layer. Table 3.3 illustrates the fact that changing the direction of drainage out of the base course from horizontal to vertically downward into a highly permeable base drainage layer increases the rates of infiltration that can be removed by 1600 to 7000 times, depending on the widths of the pavements being drained.

This table gives the rates of infiltration that can be accepted and discharged by "lateral drainage" for several "standard" roadbed materials in which no base drainage layer is provided and by "downward drainage" in which a highly permeable base drainage layer is provided. It is seen that without subdrainage these materials can discharge only minuscule rates of infiltration—not even a light drizzle—but where placed over a highly permeable base drainage layer, they can accommodate significant amounts of surface infiltration.

The classes of road building materials included in Table 3.3 are considered to be representative of some of the more permeable base and subbase materials normally used. Some highway departments have used a standard of 6 to 10 ft/day for "permeable" materials for roadbed drainage, although many of the base and subbase materials in widespread use have permeabilities much lower than these limits. Published reports examined by the author indicate that the "normal" bases for concrete pavements often contain from 10 to 20 percent of sizes finer than a no. 200 sieve. The coefficients of permeability for these materials can be expected to be in the general range of 1 ft/day to 0.10 ft/day or less. When base and subbase materials are stabilized with cement even lower coefficients of permeability can be expected. A technical manual for airport design (FAA, 1970) says that base course materials meeting FAA Construction Specifications P-208 or P-209 would have coefficients of permeability in the range of $10^{-1}$ to $10^{-4}$ cm/sec (300 to 0.3 ft/day). This manual says, "Pavement courses that have been adequately stabilized with cement or bitumen are impermeable . . . ."

Downward flow is the most efficient way to drain primary pavement layers, because of the combining effects of several beneficial factors: (1) the forces of gravity pull downward, in the direction of flow, to give the greatest possible benefits from these forces; (2) the seepage distances are the shortest possible—just the thickness of the

**TABLE 3.3** Potential Increase in Infiltration by Changing the Direction of Drainage

| Class of aggregate in base course | $k$ (ft/day) | Rate of infiltration removed (in./hour): Lateral drainage[a] (no base drainage layer) $W = 35'$ | $W = 75'$ | $W = 150'$ | Downward drainage[b] All widths |
|---|---|---|---|---|---|
| Well-graded mixture | 2 | 0.0006 | 0.0003 | 0.00014 | 1.0 |
| Concrete sand, high dust | 5 | 0.0015 | 0.0007 | 0.00035 | 2.5 |
| Concrete sand, low dust | 10 | 0.0030 | 0.0014 | 0.00070 | 5.0 |
| Well-graded filter aggregate | 20 | 0.0060 | 0.0028 | 0.00140 | 10.0 |

[a] Assumes the following: 12-in. thick bases, with half-widths = $W$; gradient = 0.02; $Q = kiA$ (layers flowing full); base is on an impermeable subbase or subgrade.
[b] Assumes an underlying base drainage layer with infinitely high permeability; no flooding of base drainage layer; free downward flow under vertical hydraulic gradient of 1.0 in the upper layer being drained.

layer(s) being drained; and (3) the discharge areas are the maximum possible—the plan area of the layer(s) being drained.

In short, downward drainage makes fullest possible use of all the factors in Darcy's law, $Q = kiA$ (see "Basic Principles" in Sec. 3.3 for discussion of Darcy's law). When water is flowing vertically downward toward a highly permeable base drainage layer, the hydraulic gradient $i$ in the layer being drained is essentially 1.0. Hence the downward discharge rate per unit area is $q = k(1.0)A = kA$; thus for a unit area, the possible rate of flow is $q = k(1.0)(1.0) = k$. Free downward flow allows a material with a coefficient of permeability of 6 ft/day to discharge 6 $ft^3$ of water per square foot per day; one with a coefficient of permeability of 10 ft/day to discharge 10 $ft^3$ per square foot per day, and so on. These are the rates of flow that are often visualized for drainage materials, but which can be achieved only when free, unobstructed vertical (downward) flow is permitted. Usually this can be assured only by providing a drainage system which includes a highly permeable drainage layer under the full width of a pavement and the necessary collector pipes and outlet pipes, designed by the methods described in this book.

## 3.3 DEVELOPMENT OF DRAINAGE CRITERIA BY RATIONAL METHODS

For many decades, groundwater hydrologists have made use of Darcy's law of flow (Darcy, 1856), also called Darcy's law, to analyze the rise and fall and spreading of groundwater mounds and the productivity or capacity of wells to supply water. Also, soils engineers have made extensive use of Darcy's law to solve a wide variety of problems in the field of soil mechanics and foundation engineering. Although these principles are applicable to the study of flow of water in pavements, they have seldom been used, until the past few years (Lovering, 1960; Cedergren, 1960; FHWA, 1973a) for this purpose.

### Basic Principles

The flow of water upward from soil formations into structural sections, downward through porous pavements into structural sections, downward into permeable subgrades, horizontally in bases, horizontally through shoulders, and so on follows fundamental laws of nature. Whenever it is possible to assign reasonable values to the controlling factors in Darcy's law, reasonable estimates can be made of seepage rates, velocities, and the like. Similarly, useful analytical solutions can often be obtained by the use of flow nets (Cedergren, 1967, pp. 148–170) for a variety of practical situations.

Darcy's law may have a number of forms, such as:

$$\begin{aligned} Q &= kiA \\ Q &= kiAt \\ Q &= v_d A \\ \frac{Q}{i} &= kA \end{aligned} \tag{3.1}$$

$$v = ki = \frac{Q}{A} = v_d$$

$$v_s = \frac{ki}{n_e} = \frac{v_d}{n_e}$$

In these expressions for Darcy's law $Q$ is the seepage quantity, $i$ the hydraulic gradient in the direction of flow, $A$ the cross-sectional area perpendicular to the direction of flow, $t$ the time, $v_d$ the *discharge velocity* (average velocity through total cross-sectional area), $v_s$ the seepage velocity (average velocity through the pore volume), and $n_e$ the *effective* porosity, or the portion of the cross section in which the water is actually flowing.

Darcy's law applies specifically for laminar flow under *nonturbulent* conditions. Under any conditions where flow is *semiturbulent* or *turbulent*, head losses are not proportional to head (as in Darcy's law) and quantities of flow and velocities of flow are less than indicated by the Darcy equations.

Many practical seepage problems can be analyzed with Darcy's law, including the flow of water into and out of pavement structural sections.

By consideration of fundamental laws of nature and by making certain assumptions (soil is homogeneous, voids are filled with water, soil is incompressible, no consolidation takes place, flow is laminar, and Darcy's law is valid), it can be shown (Terzaghi, 1943; Cedergren, 1967, pp. 98–100) that if $k$ is constant and if flow is *two-dimensional*, the following relationships will exist:

$$\frac{\partial^2 h}{\partial x^2} + \frac{\partial^2 h}{\partial z^2} = 0 \qquad (3.2)$$

Equation 3.2 is a form of the Laplace equation for two-dimensional flow. The term $h$ is head, and $x$ and $z$ are cartesian coordinates in two dimensions. A similar equation with three terms on the left side of the equals sign applies to three-dimensional flow.

Solutions to Eq. 3.2 can be obtained by graphical flow nets. Each flow net is made up of two families of curves that are basically "squares," although the sides are generally curved. One set of lines represents possible paths of flow of water, and these lines are called *stream lines* or *flow lines*. The other set represents lines of equal energy or head, and are called *equipotential lines*. Although mathematical and computerized solutions have been obtained for a number of types of problems, complex cross sections have been analyzed by graphical methods (Cedergren, 1967, Chapters 4 and 6; 1973), and a number of examples of the graphical solutions are given in this book (Figs. 3.15, 4.14, 4.16, 4.19, 4.23, 4.24, 4.26, and 4.29).

Solutions to the Laplace equations can be developed by mathematical computations (greatly aided by modern computer technology) and by various kinds of analogies and models. Childs (1946) describes the use of electric analogs to study flow to pipes and ditches. The U.S. Army Corps of Engineers (1946) describes the use of viscous fluid models (Helle-Shaw) to determine the time-drainage characteristics of pervious base courses of various depth-to-length ratios with the aid of Darcy's law. Casagrande and Shannon (1952) present the results of laboratory model tests, field tests, and approximate theoretical solutions for the analysis of seepage out of base drainage layers for airfield pavements.

Nettles and Schomaker (1967) describe laboratory seepage model tests that evaluated several factors influencing the degree of soil infiltration through pipe openings. Among the factors studied were the grain size (in relation to the size of joint openings), plasticity and density of the soil, and vibration effects. Practical criteria were obtained with the aid of these tests, showing the usefulness of models for studying seepage and drainage problems.

The author has used sand models to demonstrate to lecture groups the differences between lateral drainage out of roadbeds (the normal kind) and vertical drainage into highly permeable drainage layers (the desired kind). Figure 3.13*a* shows such a model with colored dye showing flow lines for lateral drainage out of a simulated roadbed through a shoulder. In this model the low trough in the sand (at the right) represents the bed in which the structural section would be placed. Even though clean, Ottawa sand was used, only a dripping flow develops.

Figure 3.13*b* shows another sand model in which a layer of asphalt-treated pea gravel (with an intervening, hand-placed, aggregate filter layer) extends under the bed of the structural section. Here the dye lines go vertically downward into the asphalt-coated pea-gravel drainage layer, and the rate of flow is 50 times greater than in the other model because the drainage layer causes the hydraulic gradients in the sand to be increased and the flow lines to be shortened.

When analytical solutions are developed for given physical conditions, the answers have a high degree of theoretical accuracy. In fact, when the work is properly done, the answers are *exactly repeatable*. This high degree of accuracy tends to give some investigators the belief that the answers are very precise solutions to physical problems. This conclusion can be very misleading, because the true accuracy of any analytical calculation can be no better than the data used in developing the solution. In the field of seepage and drainage, the physical data used in calculations may in many cases be only approximate. Permeabilities, for example, are often only crudely estimated. For this reason analytical or graphical solutions to seepage problems should be regarded as a means for determining the *general order of magnitude of problems and for developing appropriate kinds of solutions*. If used in this way, they can help engineers to avoid reaching completely illogical and erroneous conclusions about major types of seepage and drainage

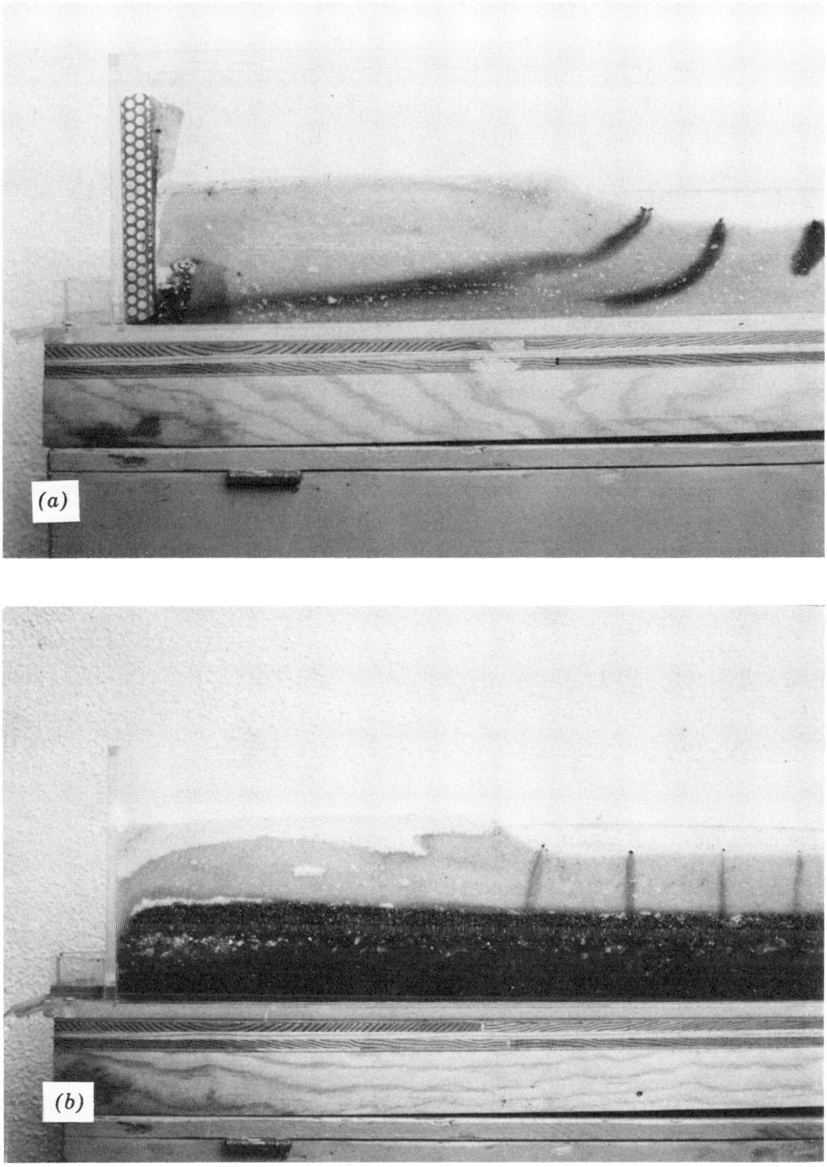


**FIG. 3.13** Model tests show up wide differences in drainability. (*a*) Lateral flow out of "normal" roadbed with no base drain layer; dye lines show lateral flow paths; outflow is a slow drip. (*b*) Downward flow into highly permeable base drain layer of Asphalt-treated pea gravel; seepage rate is increased 50 times over that in Fig. 3.13*a*.

problems. When properly used, they point up serious deficiencies in some "standard" practices in road building and drainage and show how to develop realistic solutions to the major problem of properly draining highways, airfields, and other important pavements.

Several practical methods of applying seepage principles for evaluating the needs and effectiveness of drainage systems for pavements are presented in the following paragraphs.

### The Inflow-Outflow Concept in Drainage Analysis

If pavement structural sections are *not* to become filled with excess water and are *not* to carry heavy wheel loads for significant amounts of time while in this condition, large accumulations of free water must be prevented. For this to be accomplished, the *outflow capabilities* of the subgrade-pavement systems must be at least equal to the inflows from all sources.

The "Final Report" prepared in the development of the recent FHWA *Guidelines* (FHWA, 1973, pp. 72–78) suggests that all possible inflows and all beneficial outflows be set up in a mathematical equation of the following form:

$$\sum O \geqq \sum I \tag{3.3}$$

In Eq. 3.3, $\sum I$ represents all inflow sources and $\sum O$ represents all outflow possibilities. Equation 3.3 can also be written in the following form:

$$O_E + O_S + O_P + O_R + O_D \geqq I_S + I_C + I_T + I_H \tag{3.4}$$

In Eq. 3.4, the terms have the following meaning:

$O_E$ = surface evaporation (usually negligible)
$O_S$ = loss by lateral seepage
$O_P$ = loss by subgrade percolation or drainage
$O_R$ = loss by pumping through cracks or joints
$O_D$ = water removed by subsurface drainage systems
$I_S$ = surface infiltration (often major source)
$I_C$ = capillary water from water table (usually minor)
$I_T$ = water transfer from adjacent wet areas or from underlying groundwater or springs, and so on
$I_H$ = water of hydrogenises (usually negligible)

In the past there has been a tendency to consider that $I_S$ is an unimportant source of water in roadbeds and to look for other sources such as $I_C$ and $I_H$ to try to explain the "mysterious" way in which water often gets into payments. Extensive field investigations that the author has been privileged to make for federal agencies gave him opportunities to examine important highways and airfields from coast to coast in a wide range of climatic areas. These investigations, together with observations over the past 30 to 35 years, have led to the conclusion that $I_S$ surface infiltration, largely from rainfall, is often the major if not the only significant source of water in pavements. Rainfall is an abundant source of water in many regions of the United States. Even in areas that are classified as semiarid, there is evidence of pumping and severe damage from excess water that occurs over perhaps 10 percent of the time each year.

Some investigators seem to be mystified at the way water gets into pavements in areas in which there can be no inflow from springs, high water tables, and so forth. This has led some to believe that $I_C$ and $I_H$ must be responsible. Although $I_C$ and $I_H$ can add to the water content of a subgrade, these sources are often minute in comparison with $I_S$ as sources of free water or excess water in pavements. The author has seen all the joints and cracks in pavements become filled to overflowing by rainfalls of as little as 0.02 in. and water bleeding or being pumped out of pavements for days and weeks after the last rain (Fig. 3.4*a*).

In some cases, $I_T$ can add significantly to the water entering structural sections.

When analyzing the flow of water into structural sections, an effort should be made to discover all potential sources of significance, and estimates should be made of the quantities that can enter.

In cases in which surface infiltration $I_S$ is the only major source of inflow, the following equation often expresses the important factors entering into the balance of inflows and outflows:

$$O_D \geqq I_S \tag{3.5}$$

which says that the rate of flow of water from the drainage system must be equal to or larger (potentially) than surface infiltration. Equation 3.5 applies to cases where $O_S$, $O_P$, and $O_R$ are all small enough to be neglected, which is often the case.

When designing subsurface drainage systems, it is good practice to include a factor of safety ($C$) in estimating the quantities of water that need to be removed. If this is done, Eq. 3.3 becomes:

$$\sum O \geqq C \sum I \tag{3.6}$$

And Eq. 3.5 becomes:

$$O_D \geqq CI_S \tag{3.7}$$

In order to include reasonable allowances for difficulties in realistically estimating factors such as $k$, it is felt that the factor of safety or margin of safety with respect to coefficients of permeability ought to be at least 4 or 5.

In developing criteria for the water-removing requirements of subsurface drainage systems, efforts should be made to determine realistic numerical values of coefficients of permeability for all media or layers through which water can enter structural sections. Darcy's law or flow nets can be used for making these estimates (see also Sec. 4.2). In addition, Darcy's law can be used for analyzing flow and determining the required thickness and permeability of the primary drainage layer, since transmissibility $= kA = Q/i$, with $Q$ being the estimated seepage quantity, $i$ the hydraulic gradient in the drainage layer, $A$ its thickness in feet times 1 linear foot, and $k$ the permeability of the drainage layer. The inflow-outflow concept as described in this section is one of the primary criteria that can be used in designing drainage systems.

### Condition of Continuity as a Drain-Design Concept

Using the inflow-outflow concept in the design of subsurface drainage systems, as noted in preceding paragraphs, can assure that base drainage layers will not become overloaded by the sources of water than can enter structural sections. If accumulations of water within a structural section and in all the component parts of a drainage system are to be prevented, the outflow capabilities of the system should increase in the direction of flow, starting at points of entry and progressing through the base drainage layer, collector pipes, and outlet pipes. A useful concept that provides a valuable check on the capabilities of entire drainage systems is a principle that can be called a *condition of continuity*. To illustrate, Fig. 3.14 shows the hypothetical flow paths of water entering through a pavement, flowing downward to the base drainage layer, laterally out of the base drainage layer, out of the collector pipe along the lower edge, and through an outlet pipe that removes the water to a safe point of discharge. A potential flow path in a pavement-drainage system is represented by the letters *A-B, B-C, C-D, D-E,* and *E-F* in this illustration. If the letter $Q$ is used to designate the seepage discharge *capabilities* of the various parts of the flow sequence, as estimated with Darcy's law or other analytical and experimental methods, the following equation needs to be satisfied if water is not to accumulate within any of the primary components of the system:

$$Q_{A\text{-}B} \leqq Q_{B\text{-}C} \leqq Q_{C\text{-}D} \leqq Q_{D\text{-}E} \leqq Q_{E\text{-}F} \tag{3.8}$$

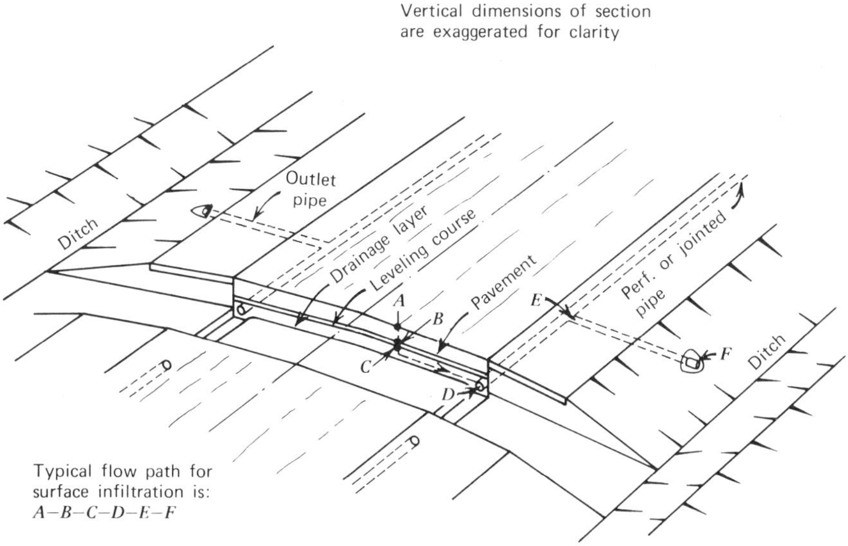


**FIG. 3.14** Illustration of flow path for "condition of continuity" in pavement drainage of surface infiltration. (Courtesy CERL.[1])

In developing a design for a subsurface drainage system Eq. 3.8 provides a useful criterion for checking on the adequacy of each of the component parts of the system. Generally, it would be desirable to apply a moderate factor of safety to the actual quantity of water expected to enter the pavement ($Q_{A\text{-}B}$) together with any other anticipated inflows, as already suggested, before establishing the flow requirements of the drainage components.

### Time Required for Water to Flow Through Pavement-Drainage Systems

In cold regions in which freezing occurs to significant depths, calculations should be made to make certain that no water can remain in structural section drainage layers long enough to freeze. This is particularly important in cold regions at times when temperatures fluctuate from above freezing in the daytime (with showers or melting of snow or ice) to below freezing at night, since cyclic fluctuations may cause the buildup of significant amounts of ice. Johnson and Lovel (1953) point out that freezing is a very serious problem, since among other things,

Reduction in load-carrying capacity is associated with a soil condition of increased moisture, decreased density, and perhaps, altered soil structure. Soils in which much water is accumulated and segregated into ice lenses during freezing will ordinarily undergo great reduction in load-carrying capacity in thawing. However, observations have revealed that significant reductions may occur with relatively small water gain and little ice segregation on freezing.

Williams (1968) points out that when pavements are constructed in cold mountain valleys, if groundwater is discharging into the structural section, conditions may be particularly susceptible to ice problems. He says, "Furthermore, if a route must cross a discharge area, recognition of the zone and an awareness of its effect on frost heaving should result in the installation of coarse fill material and an expanded subdrain network designed to alleviate conditions conducive to frost heaving."

Calculations of travel times in drainage systems can be made with Darcy's law in the form:

$$v_s = \frac{ki}{n_e} \tag{3.9}$$

using the effective coefficient of permeability $k$ of the drainage layer, the effective porosity $n_e$, and the slope $i$ in the direction of flow.

After the calculation for the possible value for $v_s$ for a given drainage layer is made, the time required for particles of water to travel the maximum distance $S$ through the layer is estimated from:

$$t = \frac{S}{v_s} \tag{3.10}$$

since the distance $S = v_s t$.

The FHWA *Guidelines* (FHWA, 1973a), suggests that for highways the maximum travel time for water in cold regions be ½ hour and in all other areas in which drainage is needed an allowable time of 1 hour be used as the criterion to prevent water from staying within structural sections long enough to cause significant structural damage under traffic impacts. For the wider, thicker pavements on airports and airfields, a maximum flow time of 1 to 2 hours is considered a reasonable range.

### Time Required to Drain After Stop of Inflows

When evaluating the total damage that could be caused by traffic on pavements containing excess water, it is important to know not only the amount of time significant inflows are occurring but also the amount of lag in the fall of saturation mounds after inflows cease. This is particularly important in relation to surface infiltration from intermittent, seasonal rainfall. If effective subsurface drains are not pro-

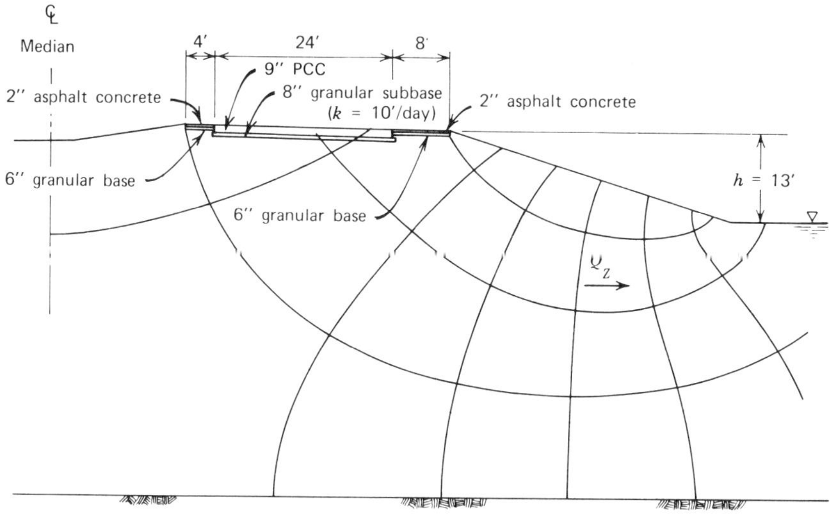


**FIG. 3.15** Flow net to determine seepage rate through subgrade (FHWA, 1973).

vided, the outflows may be largely by seepage downward into the subgrade and other minor losses. Flow nets such as the one shown in Fig. 3.15 can provide a way to estimate outflows by subgrade drainage. From a flow net, the total seepage quantity can be estimated from the equation,

$$Q = kh\frac{n_f}{n_d} \tag{3.11}$$

in which $k$ is the effective coefficient of permeability in the downward direction, $h$ is the head causing the flow, and $n_f/n_d$ is the *shape factor* of the flow net, which is the ratio of the number of *flow channels* $n_f$ to the number of *equipotential drops* $n_d$.

The rate of downward seepage can also be estimated with Darcy's law by using the average hydraulic gradient obtained by measuring the spacing between the flow net equipotential lines directly under the structural section, and then determining the hydraulic gradient $i$ from the ratio $(\Delta h)/(\Delta l)$. The *unit* seepage quantity (per square foot of pavement area) is estimated by starting with:

$$Q = kiA \tag{3.1}$$

but since $A = 1.0$, the unit seepage quantity $q$ becomes,

$$q = ki \tag{3.12}$$

To estimate the time-lag of free water mounds within structural sections after it stops raining, the volume of water contained in joints, cracks, voids, and pore spaces in pavement and base layers is estimated and designated $Q_p$. Then the time for this quantity of water to drain out by downward seepage into the subgrade is estimated from the equation:

$$t_{100} = \frac{Q_p}{q_s} \tag{3.13}$$

In Eq. 3.13 $t_{100}$ represents the time for 100 percent drainage of the quantity of water $Q_p$ by downward seepage into the subgrade at a discharge rate of $q_s$.

To illustrate the use of Eq. 3.13, assume that the subgrade of a runway or highway pavement has a coefficient of permeability (downward) of $1 \times 10^{-6}$ cm/sec (0.003 ft/day) and the average hydraulic gradient is 0.2. Also assume that the combined porosity of pavement and base totals 0.11 $\text{ft}^3/\text{ft}^2$ of pavement, for $Q_p = 0.11\ \text{ft}^3$. From Eq. 3.12, the potential drainage rate is $q_s = ki = 0.003(0.2) = 0.0006$ $\text{ft}^3$/day. Then, the time for complete drainage through the subgrade is $t_{100} = Q_p/q_s = 0.11/0.0006 = 180$ days. For 50 percent drainage, the time would be half or 90 days.

In this example, if the subgrade permeability were $1 \times 10^{-5}$ cm/sec (0.03 ft/day), the time for complete drainage would be one-tenth of that calculated, or 18 days; if the permeability were $1 \times 10^{-4}$ cm/sec (0.3 ft/day), the time would be one-hundredth of 180, or 1.8 days.

If a base drainage layer is provided with collector pipes along the lower edges, a U.S. Army Corps of Engineers (1965) formula may be used for relating the factors affecting the rate of drop in saturation in the base after it has become completely filled with water:

$$t = \frac{n_e D^2}{2880 k H_o} \tag{3.14}$$

In Eq. 3.14, $t$ is the time for 50 percent drainage of a sloping base course with a drain at its lower edge, $n_e$ is the effective porosity of the base, $D$ is its sloping width, $k$ is the coefficient of permeability of the base in feet per minute, and $H_o = H + sD$ where $H$ is the thickness of the base, and $s$ is its cross slope.

Using Eq. 3.14, it can be seen that if a value for $t$ of 0.1 day (2.4 hours) is used for a criterion for 50 percent drainage of pavements having porous bases 1.0 ft thick, on a slope of 0.01, an effective porosity of 30 percent, and a sloping width of 100 ft, the coefficient of permeability would need to be $k = n_e D^2/2880 H_o t = (0.3)(100^2)/2880(2.0)(0.1) = 5.2$ ft/minute $= 7500$ ft/day.

## 3.4 SUMMARY COMMENTS

In this chapter it has been pointed out that the present state of the art of pavement construction does not guarantee that pavements can be made watertight for extensive periods of time, and that large quantities of water can get into most pavements after moderate lengths of exposure to weather and traffic. In the past, subsurface drainage has been generally looked on as a qualitative problem with general solutions, when in reality it is a quantitative problem with definite solutions for individual situations. Trying to design subsurface drainage systems or establish their water-removing needs on the basis of rules of thumb, standard practices, or intuition can be very misleading and usually has produced systems with inadequate capabilities. The rational methods provide means for developing solutions and drainage systems that are capable of giving pavements a high level of protection against exposure to excess water. Several methods are presented to aid in designing drainage systems that will have adequate permeabilities and thicknesses to remove water rapidly and effectively from structural sections.

### References

American Concrete Institute Committee 504, ACI (1970), "Guide to Joint Sealants for Concrete Structures," *ACI Journal,* July 1970.

Barber, Edward S. (1962), Discussion, ASCE Paper No. 2623 by H. R. Cedergren, "Seepage Requirements of Filters and Pervious Bases," *Transactions*, ASCE, Vol. 127, Part I, p. 1110.

Bruce, Arthur G. (1934), *Highway Design and Construction*, International Textbook Company, Scranton, Pennsylvania, pp. 371–373.

Butterfield, Glen R. (1964), "Sand and Gravel Core at Howard A. Hanson Dam," U.S. Committee on Large Dams (USCOLD) *Newsletter*, May 1964.

Casagrande, A. and W. L. Shannon (1952), "Base Course Drainage for Airport Pavements," *Transactions,* ASCE, Vol. 117, pp. 792–820.

Cedergren, H. R. (1960), "Seepage Requirements of Filters and Pervious Bases," ASCE *Proceedings*, Journal of the Soil Mechanics and Foundation Division, Paper No. 2623, October, 1960; also, *Transactions,* ASCE. Vol. 127, Part I, pp. 1090–1113.

Cedergren, H. R. (1967), *Seepage, Drainage, and Flow Nets*, John Wiley & Sons, Inc., New York, p. 3 pp. 371, 148–170, 98–100, Chapters 4 and 6.

Cedergren, H. R. and W. R. Lovering (1968), "The Economics and Practicability of Layered Drains for Roadbeds," Highway Research *Record* No. 215, HRB, pp. 1–7.

Cedergren, H. R. (1973), "Seepage Control in Earth Dams," in *Embankment-Dam Engineering, Casagrande Volume*, John Wiley & Sons Inc., New York, 1973, pp. 21–45. (Chapters were written by former students of A. Casagrande.)

Cedergren, Harry R. (1974), "Drainage and Dewatering," Chapter 6 in *Foundation Engineering Handbook,* (edited by H. Y. Fang and Hans Winterkorn), Van Nostrand Reinhold Company, New York, 1974, in press.

Childs, E. C. (1946), "The Water Table, Equipotentials, and Streamlines in Drained Land," *Soil Science*, Vol. 62, No. 2, pp. 183–192.

Cotton, H. E. (1937), "Methods of Subgrade Drainage," *Roads and Streets*, Vol. 80, No. 5, May 1937, pp. 61–64, 66, 68, and 70.

Darcy, H. (1856), *Les fontaines publiques de la ville de Dijon.*

Federal Aviation Agency, FAA (1970), "Airport Drainage," Advisory Circular 150/5320-5B, Department of Transportation, Washington, D.C., July 1, 1970.

Federal Highway Administration, FHWA (1973), *Final Report, Studies for the Development of Guidelines for the Design of Subsurface Drainage Systems for Highway Structural Sections,* pp. 11–17; 46–66; 72–78.

Federal Highway Administration, FHWA (1973a), *Guidelines for the Design of Subsurface Drainage Systems for Highway Structural Sections.*

Indian Roads Congress (1967), "Road Drainage Practices Around the World," *Journal*, September 1967, pp. 71–138.

Izzard, Carl F. (1944), "A Rational Approach to Drainage of a Pervious Subbase," *Proceedings*, Highway Research Board, Vol. 24.

Johnson, A. W., and C. W. Lovell (1953), "Frost Action Research Needs," Highway Research Board, *Bulletin*, No. 71, pp. 99–124.

Keene, P. and S. E. Horner (1951), "Present Practice in Subsurface Drainage for Highways and Airports," Highway Research Board *Bulletin* No. 45, 1951, p. 20.

Lancaster, C. M. and W. C. Davis (1956), "Investigation to Determine the Effectiveness of Underdrains in Draining Granular Subbases Under Concrete Pavements," Missouri State Highway Commission (unpublished report).

Latter, B. V. (1970), "Concrete Joints in Airfield Pavements—a Personal View," *Engineering and Public Works Review*, August 1970, pp. 887–889.

Lovering, W. R. (1960), "Efficient Filtering for Effective Drainage," *Asphalt Institute Quarterly*, May 1960.

McClelland, Bramlette and L. E. Gregg (1944), "Methods of Analysis of Flow Problems for Highway Subdrainage," Highway Research Board *Proceedings*, 24th Annual Meeting, pp. 364–76.

Minor, Carl E. (1961), "Subsurface Drainage Problems on Multilane Freeways," WASHO Conference *Proceedings*, Materials, First Session.

Muir, Levi (1942), "Present Practice in the Drainage of Highways," Highway Research Board *Proceedings*, 22nd Annual Meeting.

Nettles, E. H. and N. B. Shomaker (1967), "Laboratory Investigations of Soil Infiltration Through Pipe Joints," Highway Research Board *Record*, No. 203, HRB.

Pole, Tudor C. G. (1970), "Joint Movement Design, and Materials," *Civil Engineering and Public Works Review*, August 1970.

Porter, Henry C. (1938), "The Preparation of Natural Soil Foundations and Soil Substructures," *Proceedings*, 18th Annual Meeting, Highway Research Board, Washington, D.C., November 28–December 2, 1938, Part II, p. 346.

Shirley, D. E. (1970), "Flexible Seals for Movement Joints in Concrete Structures," *Civil Engineering and Public Works Review*, August 1970.

Slaughter, George M. (1973), "Evaluation of Design Methods of Subsurface Drainage Facilities for Highways," Georgia Institute of Technology, GDOT Research Proj. No. 6901, "Final Report."

Terzaghi, K. (1943), *Theoretical Soil Mechanics*, John Wiley & Sons, Inc., New York, pp. 239–40.

Tulayev, A. Y. and G. D. Prasad (1965), "Design of Drainage Facilities for Pavements," Indian Road Congress, Vol. 29, No. 1, July 1965, pp. 41–74 (India and USSR).

U.S. Army Corps of Engineers (1946), *Subsurface Drainage Investigation, Report on Viscous Fluid Model Tests*, Office of the Chief of Engineers, Washington, D.C., April 1946, p. 40.

U.S. Army Corps of Engineers (1965), "Drainage and Erosion Control—Subsurface Drainage Facilities for Airfields," Technical Manual TM-5-820-2, Headquarters, Washington, D.C., August 1965, p. 10.

U.S. Army Corps of Engineers (1974), "Methodology and Effectiveness of Drainage Systems for Airfield Pavements," prepared for Construction Engineering Research Laboratory (CERL), Champaign, Illinois, Chicago District, by H. R. Cedergren.

Vallerga, B. A. (1959), "Improving Design Standards for Heavy Duty Asphalt Pavements," *Proceedings*, WASHO 38th Annual Conference, Billings, Montana, 1959.

Williams, R. E. (1968), "Ground Water Flow Streams and Related Highway Pavement Failure in Cold Mountain Valleys," *Journal of Hydrology*, Amsterdam, Netherlands, Vol. 6, No. 2, April 1968, pp. 183–93.

CHAPTER FOUR

# Estimating Inflows and Outflows

## 4.1 BASIC CONSIDERATIONS

If subsurface drainage systems are to protect pavements from harmful accumulations of free water within structural sections (including the boundaries between successive structural layers), rapid water removal must be provided for the full widths of heavily traveled portions of pavements.

The basic type of subsurface drainage system that can provide maximum protection at least overall cost consists of the following key features as illustrated in Fig. 4.1: (A) a layer of macadam-type, open-graded, highly permeable base drainage material, usually placed under the wearing course and usually substituted on an inch-for-inch basis for other base course material; (B) a subbase or other "filter" course under the macadam-type drainage layer—for structural strength and to prevent the intrusion of soil fines into the layer—and an impermeable membrane where needed to keep surface water out of sensitive subgrades; (C) a collector pipe in a shallow or deep trench, depending on the depth to groundwater and the frost penetration depth, along the lower edge of the drainage layer; (D) outlet pipes at suitable intervals to remove any excess water that gets into the system, by gravity drainage whenever possible or by pumping in the rare cases where gravity drainage is not feasible; and (E) outlet posts and markers to help maintenance personnel locate outlets and to protect the outlets from damage.

Several useful criteria for designing effective subsurface drainage systems are described in Chapter 5. Inherent in most seepage calculations for subsurface drains are the quantities of water that can get into the systems and the capabilities for removing the water that reaches them. Chapter 4 is devoted primarily to methods for estimating these inflow quantities and outflow capabilities.

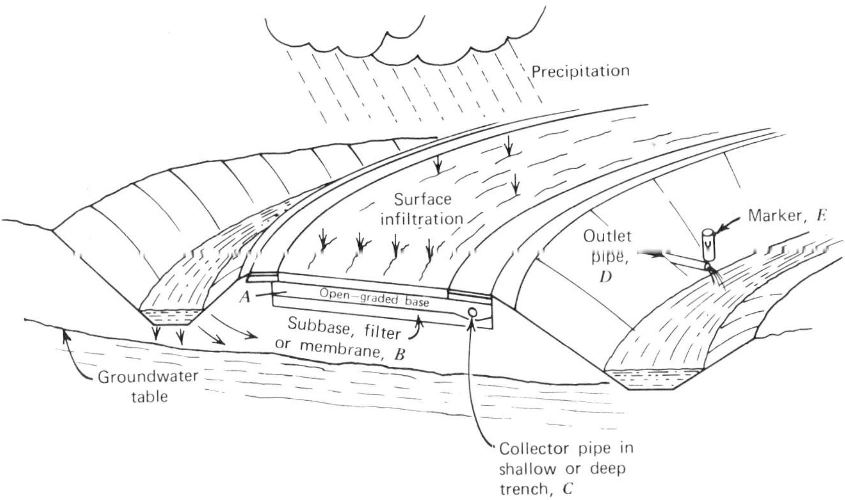


**FIG. 4.1** Basic components of subsurface drainage systems (FHWA, 1973).

## 4.2 ESTIMATING INFLOW RATES

As pointed out in Chapter 3, estimates should be made of all important sources of inflows into structural sections. In many areas surface infiltration is the major if not *only* important source, but at other locations high groundwater, spring inflows, and other localized flows will add to the inflows. In all such cases, attempts should be made to estimate the total probable inflow rates. In Eq. 3.4, the following potential sources of inflow are included:

$I_S$ = surface infiltration (often major source)
$I_C$ = capillary water from water table (usually minor)
$I_T$ = water transfer from adjacent wet areas, wet shoulders, underlying groundwater, springs, and so on (sometimes major source)
$I_H$ = water of hydrogenesis (usually negligible)

### Surface Infiltration

The amount of surface infiltration that can enter into a structural section is controlled by the lesser of: (a) the amount allowed by the effective permeability of the wearing course, or (b) the amount of supply available (design precipitation rate).

Researchers in Europe have been using the term "global permeability" ($k_g$) to represent the average or effective coefficient of permeability of a given macroscopic amount of pavement, including the effects of discontinuities such as cracks, joints, and any other areas or openings that allow water to enter—in addition to the intrinsic permeability of the pavement itself. Although the concept of "global permeability" is not new, methods for determining working values have been almost entirely lacking. If a pavement is constructed on the highly permeable, macadam type of base drainage layer illustrated in Fig. 4.1 and described in Chapter 6, with pipe outlets, the pavement-drainage system becomes, in effect, a large permeameter, because essentially all the head is used up in the pavement and base layer(s) through which water is entering. If measurements of outflows from the drains are recorded periodically during and after rainstorms, calculations can be made of the average or "global" permeability of pavement and base layer(s) for conditions existing at the time of the tests.

Another way to estimate "global" permeability of existing pavements is to make inflow tests in localized areas where water that enters has relatively free discharge under the pavements. In 1972 the author took part in such tests at airfields in the United States. In one test (see Fig. 3.2), a gallon of water entered into 5 lineal feet of construction joint in about one minute. This is an inflow of 1 gpm or 190 ft$^3$/day for a pavement area of 5 $\times$ 25 ft, with longitudinal joints every 25 ft, neglecting transverse joints and cracks. This is an average rate of 190/125 = 1.5 ft/day/ft$^2$ and the equivalent of a "global" permeability of 1.5 ft/day or 0.0005 cm/sec. For a downward hydraulic gradient of 1.0, this will allow an average inflow equivalent to a rainfall rate of 18 in./day, or 0.75 in./hour.

If dependable data become available on the effective or "global" permeabilities of various kinds of pavements, this could provide a practical way to estimate quantities of surface water infiltration that need to be removed by subsurface drainage systems. But, the amount of water that can enter into a given pavement will vary with the age and condition of that pavement; the width and spacing of cracks, joints, and so forth; and the porosity and permeability of the pavement material. Also, the amount of water that can enter depends on the diligence of maintenance and the effectiveness of the sealing operations. At present so little is known about the true permeabilities of large paved areas that the only reasonable procedure appears to be to estimate surface water infiltration on the basis of a design precipitation rate.

The Federal Highway Administration's *Guidelines* (FHWA, 1973) recommends using the 1 hour/1 year frequency precipitation rate as a design precipitation rate. This is the maximum rainfall in 1 hour that can be expected to occur on the average of one time each year. Contours of the 1 hour/1 year frequency precipitation rates for the continental United States are given on the map in Fig. 4.2. Figure 4.2 provides a beginning guide, but it must be recognized that there are many local variations that do not show on this map, and in individual areas designers should refer to local, more detailed maps or records to determine practical design precipitation rates.

The design infiltration rate used in the *Guidelines* is only one-third to two-thirds of the design precipitation rate, because normally it is expected that less than 100 percent of the water falling on a pavement will enter through the surface. To obtain the design infiltration rate, the *Guidelines* suggests that the design precipitation rate (1 hour/1 year frequency rate) be multiplied by a coefficient between .50 and .67 for portland cement concrete pavements and .33 to .50 for asphalt concrete pavements. As more information becomes available on the actual amounts of water that enter pavements, and particularly if it becomes possible to improve the watertightness of pavements, it might be possible to use somewhat lower coefficients than those suggested.

Although rainwater and other forms of surface water are very commonplace in most regions of the world, their abundance as a source of pavement infiltration has often been underrated. Photographs in Fig. 4.3 show pavements with plentiful supplies of water to enter through their surfaces. Figure 4.3*a* is of an interstate highway in the Pacific Northwest after a shower. Figure 4.3*b* is another photograph in that area while it is raining; the passing lane (in the distance) has a noticeable layer of water on its surface, although the outside lane (in the foreground) has had its surface swept by wind created by fast-moving vehicles in larger numbers and shows less water. Figure 4.3*c* shows a state highway, in an eastern state on a shallow fill several days after a light rain. Figure 4.3*d* is of an airfield taxiway (widening near a runway) in a semiarid climate in which up to ½ in. of water is still standing on the pavement several hours after a moderate rain.

Since the most minimal drainage system designed according to the methods outlined in this book will have hundreds to thousands of times the water-removing capabilities of the normal, slow-draining pavements, it is felt that designers have a good deal of latitude in selecting design infiltration rates (the 1 hour/1 year frequency rate multiplied by an appropriate coefficient). When large discharge capabilities are

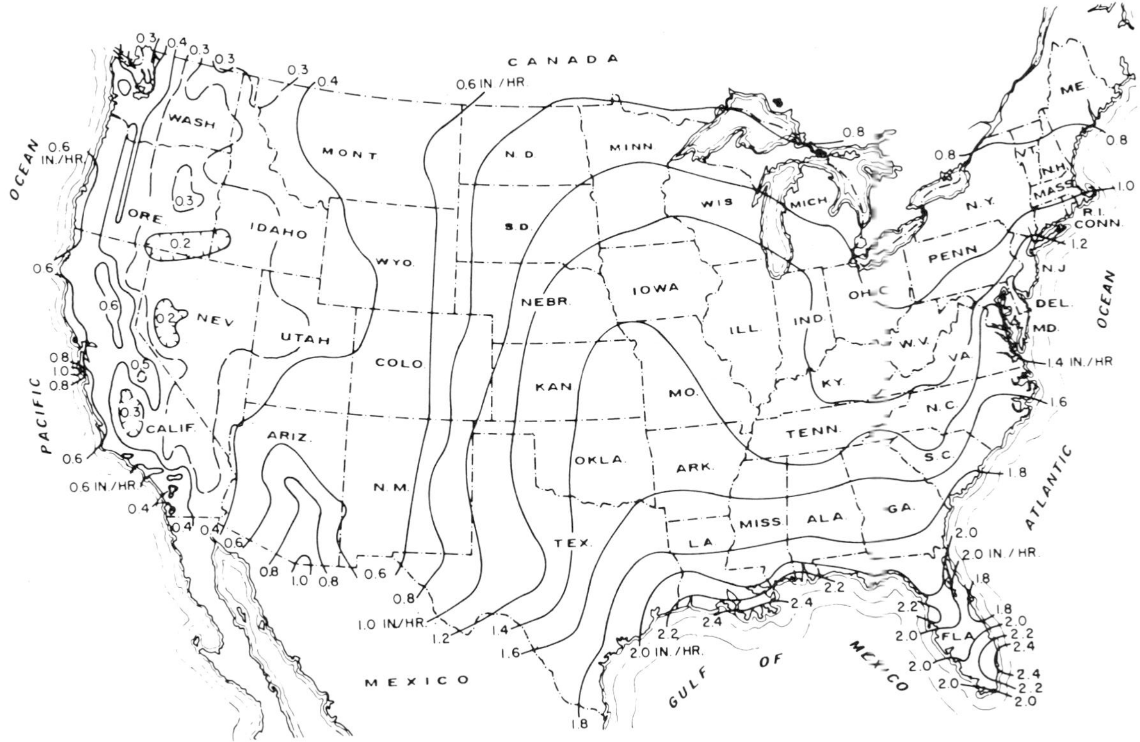


FIG. 4.2 The 1 hour/1 year frequency precipitation rates in the United States (FHWA, 1973).

**FIG. 4.3** Large amounts of surface water are frequently available to infiltrate pavements. (*a*) Much water on AC pavements of interstate highway in Pacific northwest; shoulder patches add to problems by trapping water; October 1970. (*b*) Another interstate highway in northwestern state, with large amount of water available to enter structural section; October 1970. (*c*) Badly deteriorated, 6-year-old AC pavement in an eastern state; much water in evidence several days after a rain; November 1970. (*d*) Up to ½-in. of water on airfield pavement a half day after a shower; West Coast state; December 1972. (Courtesy CERL.[1])

FIG. 4.3 (Continued).

needed, they can be obtained by selecting relatively coarse aggregate sizes for the drainage layer (see "Drainage Materials," Sec. 6.2).

### Groundwater Inflows

*General Control over Groundwater.* In areas of high groundwater levels, excessive sidehill seepage, or spring inflows, a number of kinds of drains can be used for minimizing the quantities of water that can reach structural sections and for general protection of highways and airfields. A number of widely distributed manuals contain discussions of methods for controlling these inflows (e.g., Armco, 1955; American Iron and Steel Institute, 1967). Also, several textbooks (e.g. Road Research Laboratory, 1954; Cedergren, 1967) have chapters or sections dealing with methods for controlling groundwater. *When systems for control of groundwater are designed, estimates should be made of the probable inflows, and designs should be developed and drainage materials selected according to the inflow-outflow concepts presented in this book.* Field investigations made to establish the locations where springs and other groundwater inflows can be expected should be continued into the wettest times of the year, since construction is usually carried out in the driest periods of the year when it is easy to overlook critical locations needing to be drained. Some of the kinds of subsurface drains that are often used for control of groundwater and sidehill seepage are shown in the following illustrations.

Figure 4.4 shows longitudinal drains for several types of roadbed conditions and groundwater levels needing control. In Fig. 4.4*a* a longitudinal drain along the left edge of the structural section is provided for control over sidehill seepage. In Fig. 4.4*b* a narrow road below the elevation of the normal water table requires fairly deep longitudinal drains along both sides to ensure adequate groundwater control. In Fig. 4.4*c* a wide pavement needs three lines of longitudinal drains to prevent groundwater from rising into the structural section. If there is a high saturation level that could endanger the stability of a cut slope, a drained buttress fill or a horizontal drain system is often provided, as shown in Fig. 6.1.

Figure 4.5 shows a portion of a long, sustained grade of a highway in hilly terrain. Here cross drains are placed at intervals to prevent seepage and surface water infiltration from accumulating in larger quantities than can be accommodated by the base drainage layer. This is a relatively new concept, used primarily within the past few years (Lovering and Cedergren, 1962).

Figure 4.6 depicts a transverse interceptor drain located at the lower end of a cut having groundwater inflows. This drain prevents seepage that enters the structural section from flowing out on the fill.

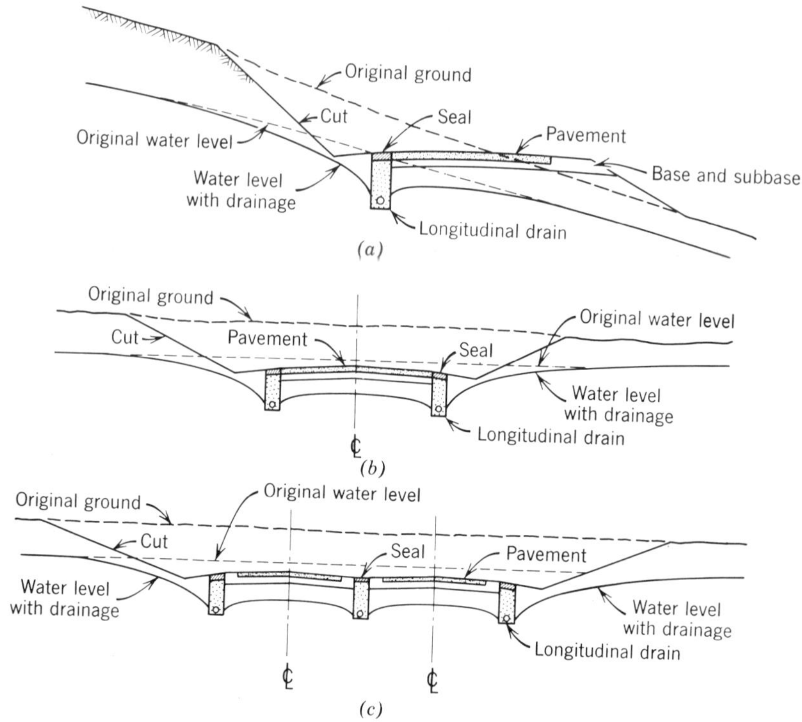


**FIG. 4.4** Longitudinal highway drains. (*a*) Side hill construction. (*b*) Narrow road in flat terrain. (*c*) Wide road in flat terrain. (*Seepage, Drainage, and Flow Nets,* John Wiley & Sons, New York, 1967, p. 365).

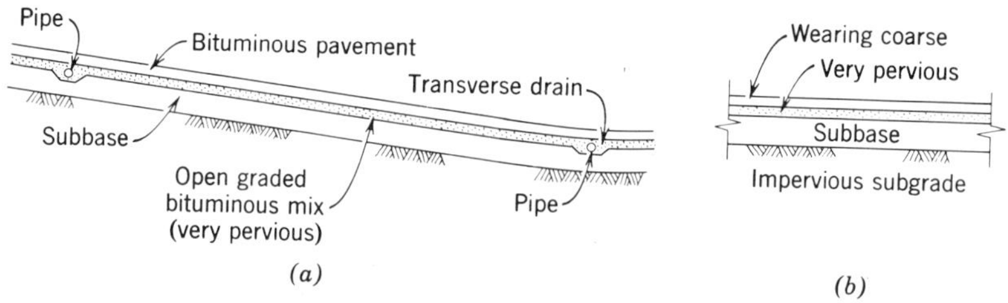


**FIG. 4.5** Pavement design providing permeable shallow layer for drainage of surface water (primarily) on long, sustained grade. (*a*) Profile. (*b*) Section. (*Seepage, Drainage, and Flow Nets,* John Wiley & Sons, New York, 1967, p. 361).

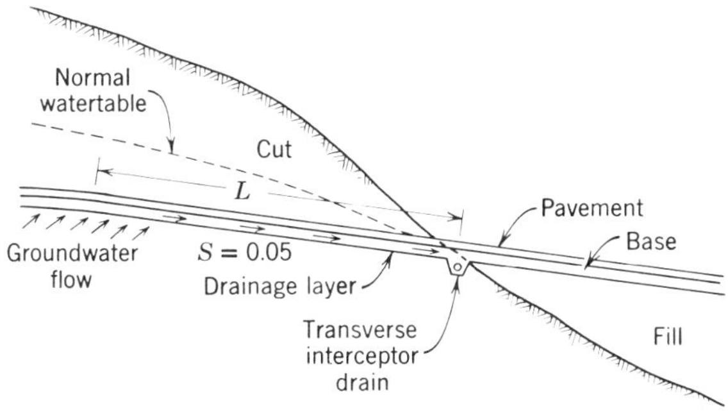


**FIG. 4.6** Transverse interceptor drain for highway. (*Seepage, Drainage, and Flow Nets,* John Wiley & Sons, New York, 1967, p. 369).

Some of the most serious problems with groundwater are those that are caused where fills are placed in such a way as to block natural flows of water out of the soil or earth formations. Figure 4.7 shows some of the kinds of localized treatments that are used to keep spring inflows or other localized groundwater flows from saturating embankments. Figure 4.8 shows a filter blanket placed over a wet foundation that might cause extensive saturation beyond the capabilities of localized treatments, such as are shown in Fig. 4.7.

Where embankments are constructed on sloping terrain in which there is extensive sidehill seepage and soft, weak strata that might not have sufficient strength to support the fill in the presence of water, a "stabilization" trench is sometimes used, as illustrated in Fig. 4.9. The back slope and the bottom of the trench are covered with a blanket of suitable drainage material that can conduct the groundwater to pipes that remove it to safe discharge points. In the past drainage materials used in stabilization trenches have often been blends of sand and gravel with coefficients of permeability in the order of 10 to 20 ft/day. If any appreciable inflows from highly permeable layers or seams are likely to enter a drainage blanket, these blended materials may not have adequate capacity. Assume, for example, that in Fig. 4.9 an inflow of 10 gpm will enter the drainage blanket for each linear foot of roadway. The minimum required coefficient of permeability can be estimated with Darcy's law as follows: A rate of 10 gpm is equal to 14, 000 gal/day or 1900 $ft^3$/day. The hydraulic gradient in the sloping blanket is approximately $1/1.25 = 0.80$. If the blanket is 2 ft thick, the required permeability $k = Q/iA = 1900/(0.8)(2.0) = 1200$ ft/day. This would require the use of a clean, coarse type of material such as pea gravel or a gravel mixture ranging in size from about ⅛ in. up to ½

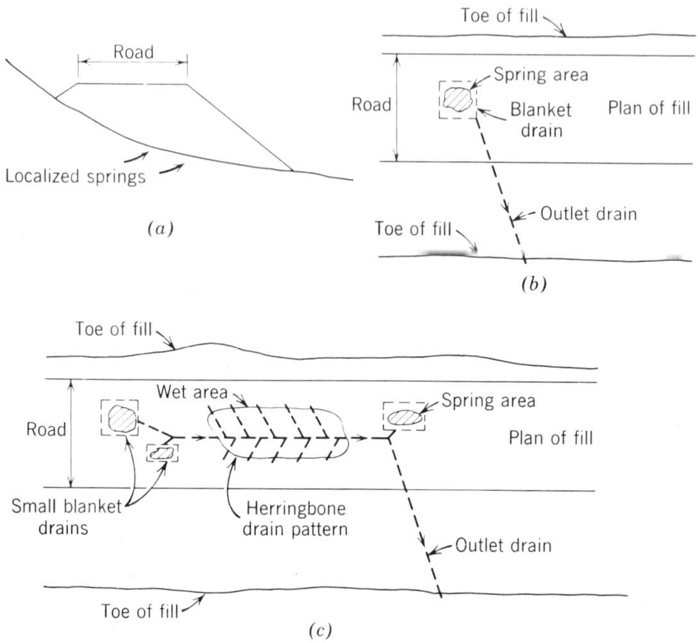


**FIG. 4.7** Localized surface drains. (*a*) Cross section of fill. (*b*) Draining a single spring. (*c*) Draining a group of springs. (*Seepage, Drainage, and Flow Nets,* John Wiley & Sons, New York, 1967, p. 280).

in. or ¾ in. maximum size. Materials of this type usually have coefficients of permeability in the order of 3000 to 5000 ft/day or higher. In this example, if it is assumed that an available material of this type will have a coefficient of permeability of at least 6000 ft/day, economies could be achieved by cutting the thickness of the blanket to 1.0 or 1.5 ft, because a thickness of 1.0 ft would have the capability of removing $Q = kiA = (6000)(0.8)(1.0) = 4800$ ft$^3$/day, which is four

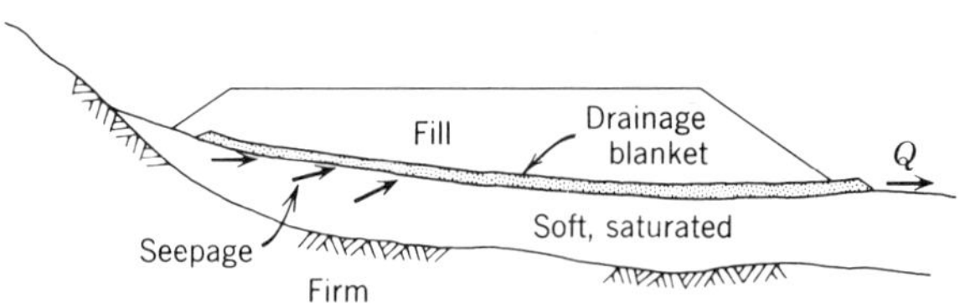


**FIG. 4.8** Drainage blanket under fill on wet foundation. (*Seepage, Drainage, and Flow Nets,* John Wiley & Sons, New York, 1967, p. 278).

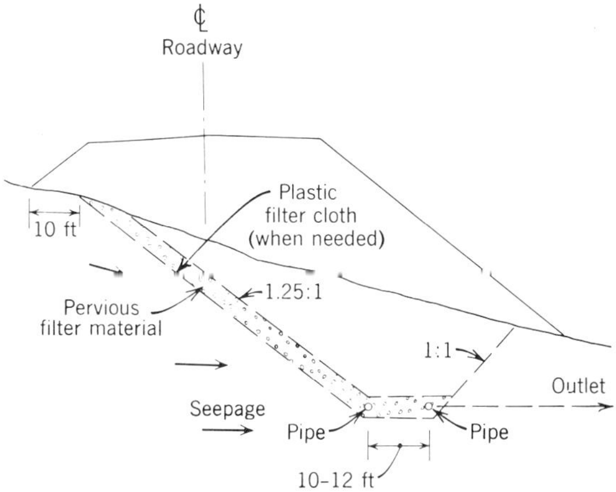


**FIG. 4.9** Cross section through a stabilization-drainage trench used for highways in wet, unstable terrain. (*Seepage, Drainage, and Flow Nets,* John Wiley & Sons, New York, 1967, p. 277).

times the estimated inflow. A reasonable amount of reserve transmissibility should always be provided in drainage layers such as this one, to allow for some intermingling of foundation soil or embankment material placed over the blanket.

When clean, coarse drainage materials are placed on soft, erodible soil formations, it is necessary to make sure that no large amounts of the soil can wash into the drainage material (see Sec. 5.5). In cases such as Fig. 4.9, it would be difficult to place a layer of fine filter aggregate to protect the coarse layer, but plastic filter cloths have been used successfully for this purpose.

Figure 4.10 illustrates some of the kinds of subsurface drains that have been used for airport drainage. This illustration was obtained from a Civil Aeronautics Administration manual (CAA, 1956). Details showing methods for controlling various kinds of groundwater conditions for airfield pavements, as recommended by the Federal Aviation Administration, are given in Fig. 4.11 (FAA, 1970).

Probably the biggest builder of air bases and airfields is the U.S. Department of the Army. According to a Department of the Army Technical manual (U.S. Army Corps of Engineers, 1965), subsurface drainage facilities may be needed for three main purposes: (a) base drainage, (b) subgrade drainage, or (c) intercepting drainage.

Base drainage is required by the army where frost action occurs in the subgrade beneath airfield pavements. Subsurface drain pipes are

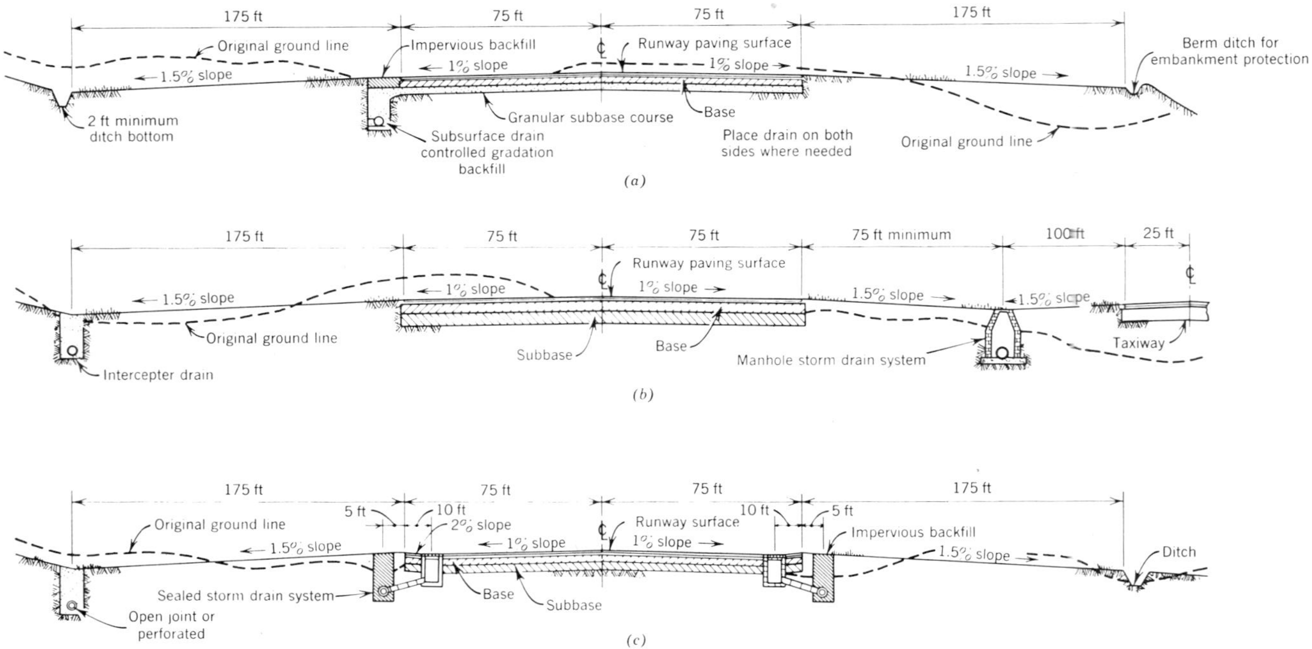


**FIG. 4.10** Typical runway cross sections. (*a*) Pavement base or subsurface drains (*b*) Landing strip drainage. (*c*) Pavement gutter drain. (Civil Aeronautics Administration).

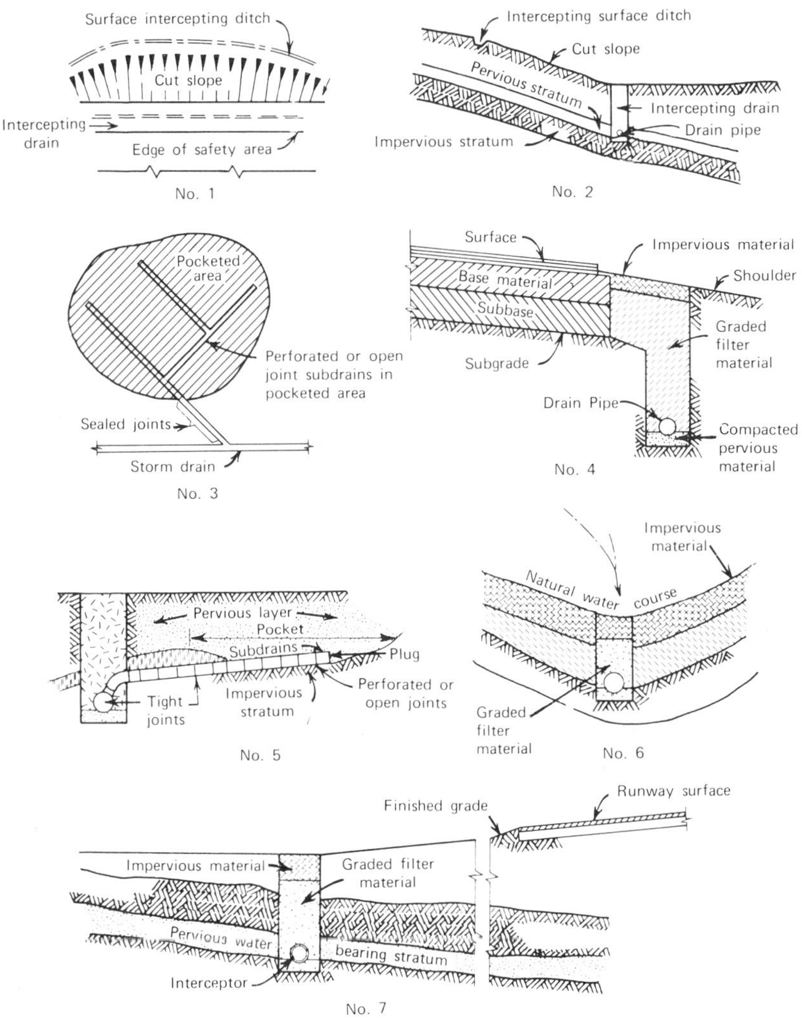


**FIG. 4.11** Types of subsurface drains for airfield pavements. (Federal Aviation Agency).

usually laid parallel and adjacent to pavement edges with pervious material connecting the pipes with the base. A formula is given in the manual that allows determination of a coefficient of permeability of the base that will assure that such a layer will be at least 50 percent drained in 10 days after it becomes completely filled with water.

Subgrade drainage is required at locations where seasonal fluctuations of groundwater may rise in the subgrade beneath a paved area to less than 1 ft below the bottom of the base course. These drains usually consist of either a system of subsurface drain trenches with pipes or a system of open ditches. The type, location, depth, and spacing of drains are determined on the basis of soil characteristics and the depth to the groundwater table. Guides are given in Corps of Engineers manuals for designing these drains.

Intercepting drainage is provided to intercept groundwater flowing in a pervious shallow stratum toward a paved area or into the face of a cut. The type and depth of drain are determined on the basis of soil type and groundwater condition.

Most highway and airport design departments have guides or manuals for designing drains for groundwater and other inflows, often using the same general approach as that used by the army in designing subsurface drains for airfields. The intent is to provide trenches or other drains of such depths and dimensions as to intercept all known sources of groundwater inflows or any that are indicated as likely from subsurface investigations and readings from piezometers, observation wells, and general observations. Where adequate drains are not provided, water can often be seen bleeding out of joints (Fig. 4.12) or from cracks in highway and airfield pavements.

*Estimating Inflows.* Any groundwater or other sources of inflow that can be expected to reach pavement structural sections should be added to surface infiltration in determining quantities needing to be removed by subsurface drainage systems for pavements (see Example 3, Section 6.3). As previously noted, field investigations of groundwater conditions should include observations and measurements of water levels in drill holes (wells), evidences of springs, and so forth, *during the wettest times of the year*; otherwise the extent of the areas needing drainage and the volumes of inflow can be grossly underestimated. Analytical solutions (flow nets, Darcy's law, and so forth) can be helpful in analyzing probable inflow quantities from groundwater and springs and in developing designs for drainage trenches, drainage blankets, and so forth. Whenever possible the inflows should be expressed in the equivalent infiltration rate in inches per hour and added to the design infiltration rate (see "Surface Infiltration" in Sec. 4.2) to obtain the total inflow quantity to be removed by the drainage system.

If it is expected that groundwater will be flowing upward into a structural section from an underlying permeable aquifer, the best possible estimate should be made of effective soil permeability and the probable hydraulic gradient causing water to flow into the roadbed.

**FIG. 4.12** Perennial bleeding at this location on an interstate highway in a central state; is fed by spring inflows from nearby cut; fall 1970.

Then with Darcy's law, the rates of inflow per square foot of roadbed are estimated to be the product of permeability and hydraulic gradient, or $q = ki$. Figure 4.13 is a chart giving the equivalent infiltration per square foot of roadbed for a range of subgrade permeabilities and hydraulic gradients. If soil permeabilities are relatively low (generally under about 0.1 ft/day or $3 \times 10^{-5}$ cm/sec), the inflowing quantities will usually be insignificant in relation to normal surface infiltration rates; however, if permeabilities are high, the inflowing seepage can be significant. Thus if the soil has a coefficient of permeability of 0.1 ft/day and the hydraulic gradient is 0.4, the equivalent infiltration $I_g$ is 0.02 in./hour (point *A*, Fig. 4.13), and if the design infiltration rate (from surface water) is 1.0 in./hour, the combined rate would be 1.0 + 0.02 = 1.02 in./hour. In such a case, the groundwater inflow is not significant. But assume that the subgrade has a coefficient of permeability of 10 ft/day and the hydraulic gradient is 0.5 (point *B*, Fig. 4.13). Now, the equivalent infiltration rate (from groundwater) is 2.5 in./hour, and the combined rate would be 3.5 in./hour. If the subgrade has a coefficient of permeability of 100 ft/day and the hydraulic gradient is 0.2 (point *C*, Fig. 4.13), the equivalent infiltration rate (from groundwater) is 10 in./hour, which is a very large rate. In this

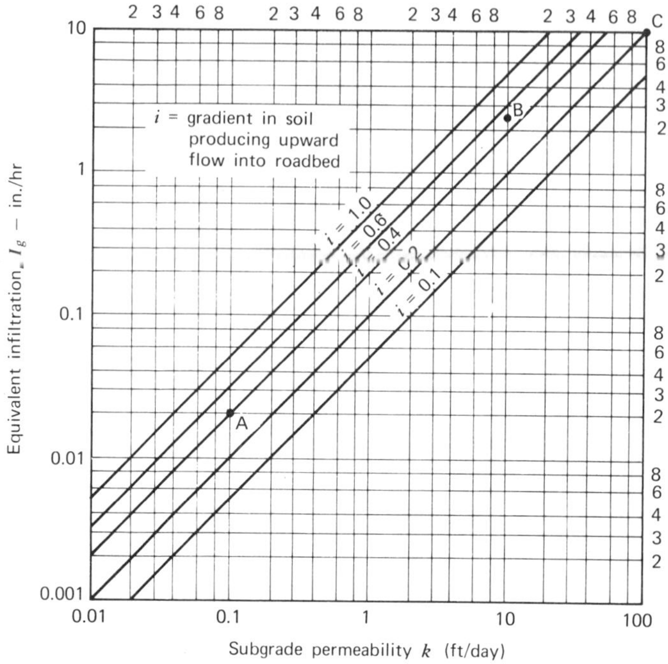


**FIG. 4.13** Chart for vertical groundwater seepage into horizontal drainage blankets from underlying artesian aquifer.

case, the combined rate (surface water + groundwater) is 11 in./hour, which would require a very high capacity drainage system.

Flow nets are also useful for obtaining solutions to various kinds of groundwater conditions. Several examples are given next.

**Example 1 Groundwater Flow into Drainage Blanket.** This flow net solution assumes that groundwater is seeping vertically upward into a roadbed from an underlying, artesian aquifer (Fig. 4.14). In many cases where freeways, expressways, or undercrossings are constructed in depressed sections below the water table (as in this example), the groundwater flows should be added to any surface infiltration to determine the total discharge requirements of the drainage system. The flow nets in Fig. 4.14 were developed to show the way that water can flow upward through a soil and then laterally in drainage blankets (base drainage layers) to enter longitudinal trench drains. In this example it is assumed that a gravel aquifer with an excess head (artesian uplift head) of $0.2D$ lies at a depth of $0.5D$ below the base drainage layer. The term $D$ is the spacing between the trench drains. For these assumptions, the average upward hydraulic gradient in the soil is 0.4, and the flow patterns take the shapes shown in these flow nets. The height of rise of saturation in the base drainage layer (with no surface water infiltration) is $h$, which depends on the relative coefficients of permeability of the soil and the drainage layer.

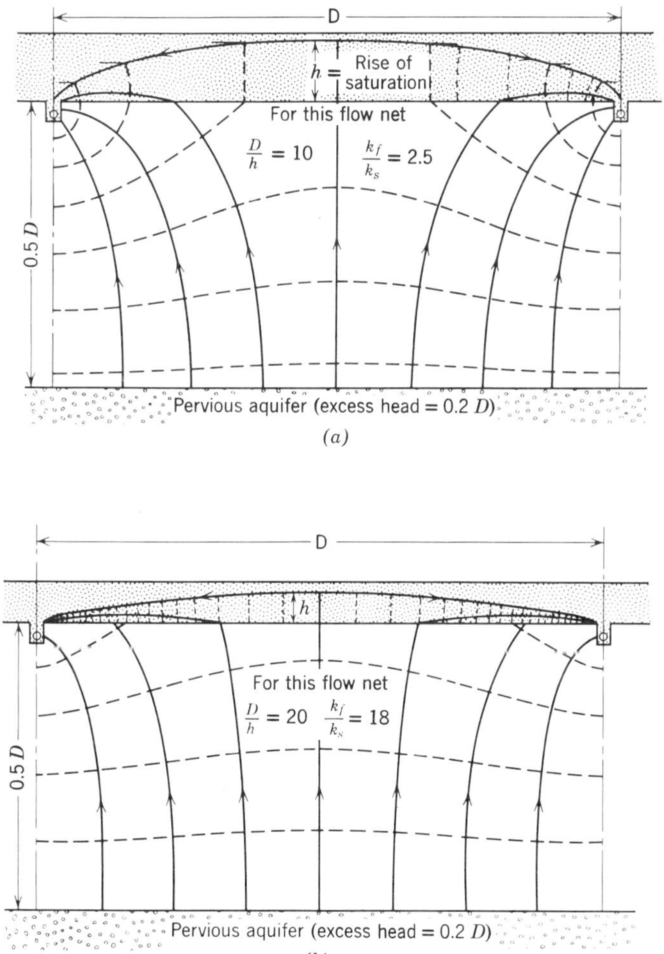


**FIG. 4.14** Typical flow nets for vertical seepage into horizontal drainage blankets from underlying aquifer. (*Seepage, Drainage, and Flow Nets,* John Wiley and Sons, New York, 1967, p. 197.)

By constructing a number of flow nets such as are given in Fig. 4.14 the curve in Fig. 4.15 was obtained. From Fig. 4.15 it is possible to determine the thickness of a base or filter of any given permeability $k_f$, in relation to the subgrade permeability $k_s$, needed to prevent saturation from rising to the top of the drainage blanket (base drainage layer). For example, assume that $D = 80$ ft, and the base drainage layer has a coefficient of permeability 1000 times that of the subgrade ($k_f/k_s = 1000$). From Fig. 4.15, the ratio $D/h$ is read as 108. Then, since $D = 80$ ft, the height of saturation $h = D/108 = 0.7$ ft. If the ratio $k_f/k_s$ is equal to 10,000, $D/h = 310$, and $h = D/310 = 0.26$ ft. For this type of groundwater inflow, if no rainwater or other infiltration needs to be added, the base drainage layer would need to be approximately 1.0 ft thick for $k_f/k_s = 1000$, or 0.4 ft thick for $k_f/k_s = 10{,}000$. If greater total inflows were expected for any reason, this would need to be considered in developing the design.

In preparing designs for a subsurface drain, it is usually advisable to consider materials of several degrees of permeability for the drainage layer. Frequently, by utilizing a higher permeability material that is readily available at about the same unit cost as another lower permeability material (of a suitable type), greater economy can be achieved, since less total thickness of high-permeability material is needed, as illustrated in this example. When highly permeable open-graded drainage materials are used, filter layers should always be provided where needed (see Fig. 5.9).

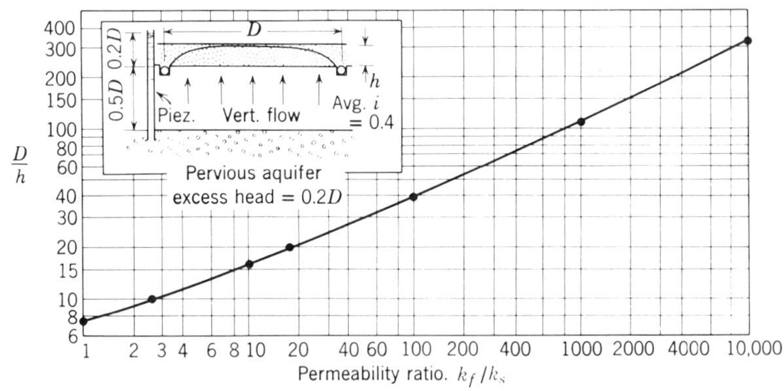


**FIG. 4.15** Chart of flow net solutions for vertical seepage into horizontal drainage blankets from underlying aquifer. (*Seepage, Drainage, and Flow Nets,* John Wiley and Sons, New York, 1967, p. 198.)

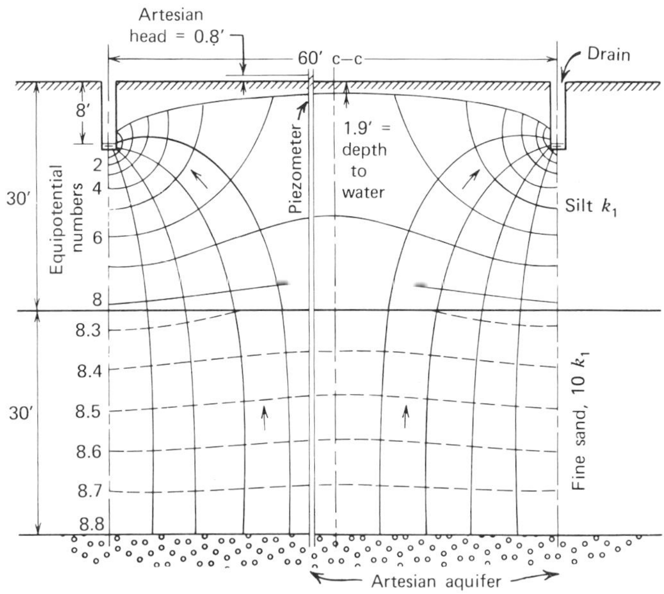


**FIG. 4.16** Flow net for vertical groundwater flow into trench drains. (*The American City,* October, 1970). (Copyright, Buttenheim Publishing Corporation, 1970).

**Example 2 Groundwater Flow into Trench Drains.** By utilizing highly permeable drainage blankets with shallow collector drains, as in Example 1, it is often possible to obtain good control over subsurface water inflows. Some designers use drainage trenches to control upward flowing groundwater. Example 2 gives a flow net solution for seepage into fairly deep trench drains used for this purpose. The typical flow net in Fig. 4.16 (several were constructed) was developed for the subsurface conditions shown—a buried artesian layer of sandy gravel 60 ft below the surface. The lower 30 ft of the soil is assumed to have a coefficient of permeability 10 times that of the upper 30 ft. Trench drains are in longitudinal lines spaced 60 ft apart. An artesian head of 0.8 ft above the ground is assumed in the aquifer. If the saturation level is to be kept approximately 2 ft below the ground level, the drain pipes would need to be 8 ft below the ground level, and the trenches would need to be about 9 ft deep. Similar solutions can be developed for any soil profiles, groundwater conditions, and so forth for which this type of drainage is planned. The chart in Fig. 4.17 was obtained from a number of such flow nets. From this chart, one can determine the amount of groundwater lowering that could be obtained by drain pipes 8 ft deep, 60 ft c-c, for various amounts of artesian head in

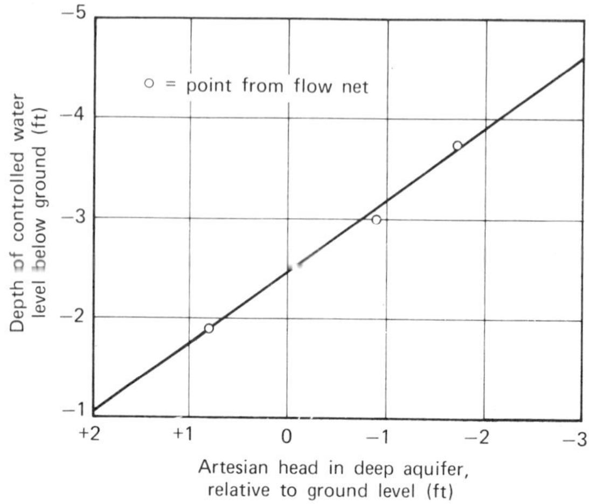


**FIG. 4.17** Depth of controlled water level below ground versus artesian head in deep aquifer—8-ft-deep drains at 60-ft c-c. (*The American City,* October, 1970). (Copyright, Buttenheim Publishing Corporation, 1970).

the buried aquifer. This could serve as a guide for designing trench drains for this kind of seepage condition.

**Example 3 Localized Spring Inflow.** In many mountainous areas seepage inflows and water entering through joints and fractures in rocks often produces local "springs" or "concentrated seeps." Figure 4.18 shows such an area where water is emerging at a number of places in a state highway in wet, hilly terrain in a western state. When water enters from localized "point" sources such as these, it may actually bulge the pavement upward if it cannot escape through cracks, and so forth, as in this photo. To analyze such a case, the flow net in Fig. 4.19 was constructed for flow in a base drainage layer above a concentrated spring inflow. It is assumed that water enters through a small "window" with a diameter $d = 12$ in. The base drainage layer is assumed to have a thickness $t = 0.5$ ft and the overlying structural layers have a total additional thickness of 1 ft. It is also assumed that an artesian head of 0.5 ft is the maximum allowable within the base at the point of the inflow (maximum head). Assuming that water flows away from the point source with equal ease in all directions (pavement is essentially flat) with an uplift head of 0.125 ft at equipotential line 1, the excess head at the point of inflow will be 0.5 ft. This is 4 equipotentials at 0.125 ft each, which is the amount that was set as the maximum allowable and is the basic control.

Using the flow net in Fig. 4.19, it is possible to estimate the minimum coefficient of permeability of a base in relation to a spring inflow quantity $q$, which can be accommodated by the base with an uplift head not exceeding 0.5 ft.

From the flow net, inflows are related to the other factors by the equation:

$$q = kh \frac{n_f}{n_d} t \tag{4.1}$$

and

$$q = 0.5k \frac{8}{4} t$$

So

$$q = 1.0kt$$

For $t = 0.5$ ft, as assumed in this example, the permeabilities in Table 4.1 would be required for a range of spring inflows.

According to the flow net solution summarized in Table 4.1, even minute spring inflows can overload blends of sand and gravel that are

**FIG. 4.18** "Spring" inflows coming out of an AC pavement in west coast area with heavy rainfall; April 1971.

normally used in bases and drainage layers under pavements. If spring inflows of several gallons a minute or more can be expected, it will be necessary to install individual drains at the springs (if they can be located during construction), or to use a blanket of coarse material with a coefficient of permeability of at least several thousand feet a day.

Many highways, such as the one shown in Fig. 4.18, are in very wet areas with much groundwater in the rainy season, but many of the wet areas dry up in the summer and do not show up during the construction. In such cases, underdrains that are called for on the plans are sometimes

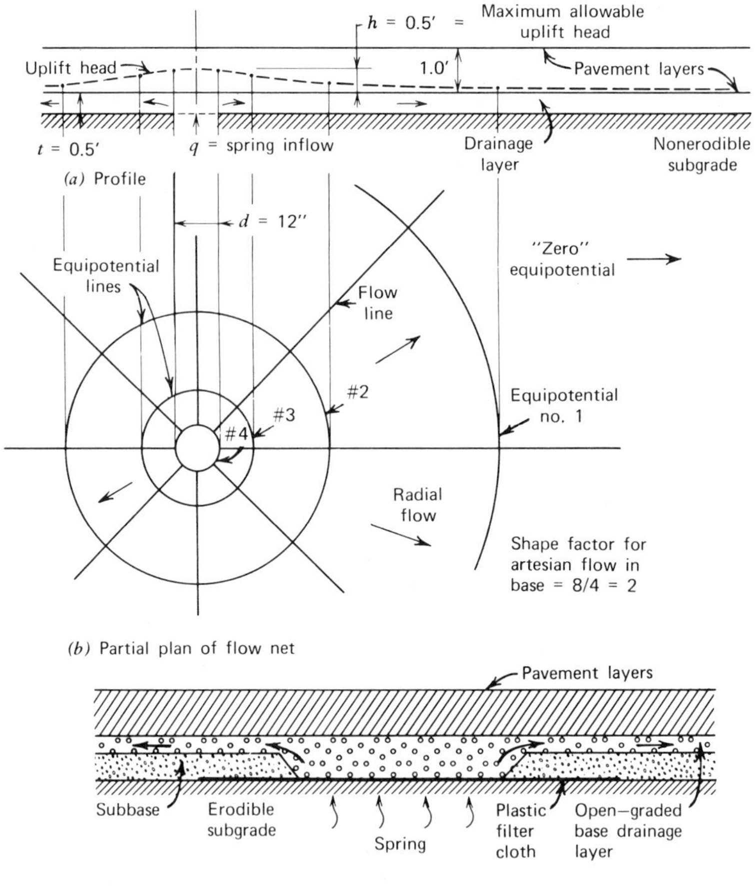


(c) Treatment at springs in erodible subgrades

**FIG. 4.19** Flow net solution for discharge of spring inflow by a "permeable" base.

**TABLE 4.1** Required Permeability of Base versus Spring Inflow (for flow net in Fig. 4.19*b*)

| Required base permeability ft/day | Spring inflow rate | |
|---|---|---|
| | $ft^3$/day | gpm |
| 4 | 2 | 0.01 |
| 40 | 20 | 0.1 |
| 400 | 200 | 1.0 |
| 4,000 | 2000 | 10.5 |

omitted because there is no evidence of water when the work is being carried out. When this is done, water often causes considerable damage to the completed pavement, and is very obvious during wet periods, as in Fig. 4.18. Provision of a drainage system with high-permeability open-graded drainage material under the full width of a pavement can protect it from inflows such as are occurring here.

Wherever coarse macadam types of bases are used, as shown here, they must be protected with a suitable subbase or filter layer meeting acceptable filter criteria (see Sec. 5.5) that will allow the required inflows to occur without permitting soil to wash into the base. If such a filter is placed over a concentrated flow producing a large amount of water, it may be difficult to obtain sufficient permeability to prevent restriction of the inflow. Where this cannot be accomplished within the restricted dimensions as shown in Fig. 4.19*a*, it may be necessary to overexcavate to a depth of several feet to enlarge the size of the hole backfilled with filter material and drainage aggregate. If spring inflows from erodible types of subgrades are anticipated, it would be advisable to have on hand a supply of an approved plastic filter cloth material and to treat these areas by placing the filter cloth on the subgrade and increasing the thickness of the base drainage layer as shown in Fig. 4.19*c*.

## 4.3 ESTIMATING OUTFLOW RATES

### General Discussion

Equation 3.4 includes the following possible ways that water can get out of pavement structural sections:

$O_E$ = surface evaporation
$O_S$ = loss by lateral seepage
$O_P$ = loss by subgrade percolation or drainage
$O_R$ = loss by pumping through cracks or joints
$O_D$ = water removed by subsurface drainage systems

When pavements do not have effective subsurface drainage systems, the water that gets in from all important sources may keep the structural sections in an essentially flooded state for many days or even weeks or months after significant rainfall events. During this time, minute quantities may be lost by $O_E$ surface evaporation, although it is believed unlikely that this factor is of any significance in most cases. Also, some water is often helped out of the joints and cracks by the pressures of traffic, the forces of gravity, and temperature changes ($O_R$ in the above list). Also, over periods of time some water is drained out by downward seepage into the subgrade $O_P$, although this factor is also relatively small for many of the compacted subgrades throughout most of the United States. Some areas have sandy subgrades that are permeable enough to provide good drainage in warm weather, but do not drain fast enough to prevent freezing in cold weather. In the spring water from melting ice or rainwater becomes trapped in the structural section and the top of the subgrade by the unmelted ice beneath it. This is the spring "break up" that can severely damage pavements.

Altogether the losses of water from pavements by $O_E$, $O_S$, $O_P$, and $O_R$ are in many cases very slow processes. Methods for estimating $O_S$ and $O_P$ are shown in this section. An indication of possible amounts of time needed for surface infiltration to drain out of "conventional" pavements, as compared with a well-drained pavement is given in Table 4.2.

The data in Table 4.2 pertain to case study sites investigated in developing the FHWA *Guidelines* (FHWA, 1973). Eight of the sites are on state highways or interstate highways where excess water (from surface infiltration) was contributing to premature deterioration of these pavements, all of which were constructed with the normally used slow-draining types of bases and subbases except the Michigan site, which had good drainage except during periods of cold weather when its sandy subgrade was frozen. Figure 4.20 shows the deep cracking that

**TABLE 4.2** Estimated Length of Time Structural Section Remains Essentially Saturated after It Stops Raining

| Case study | Estimated time (days) | Relative time (Eureka = 1) | Special notes |
|---|---|---|---|
| California | 20 | 2000 | Section is on 50 to 60 ft high clayey sand embankment; slow drainage; no groundwater |
| Connecticut | 15 | 1500 | Section is on low side of superelevated curve; clayey subgrade layers over sandy fill; poor drainage; no groundwater |
| Eureka (Humboldt Co. Calif.) | 0.01 | 1 | Highly permeable drainage layer under full width of traveled way with outlet pipe; heavy rainfall and sidehill seepage |
| Georgia | 12 | 1200 | Section is on clayey sand-silty sand fill; crushed aggregate subbase is blocked by soil on slope |
| Michigan | 0.2 (unfrozen)<br>Infinite (frozen) | 20<br>Infinite | Section is on pervious sand fill; subbase is daylighted; subbase and top of subgrade become frozen at times during winter |
| Oklahoma | 15 | 1500 | Has "black base" which is not directly drained; section is on clayey sand fill; no groundwater |
| Pennsylvania | 12 | 1200 | Section is on silty sand fill; has 6-in. diameter underdrain pipe under subbase, which is also daylighted; no groundwater |
| Utah | 8 | 800 | Section is on clayey sand fill; base course is daylighted; slow drainage; no groundwater |
| Washington | 5 | 500 | Section is on high side of superelevated curve; has 2-in.-thick porous subbase which has no pipes and is not daylighted; is on silty sand-clayey sand fill |

From Table 5, *Final Report, Studies for the Development of "Guidelines for the Design of Subsurface Drainage Systems for Highway Pavement Structural Sections,"* Federal Highway Administration, Washington, D.C., 1973.

**FIG. 4.20** Test hole in badly broken AC on CTB on state highway in Pacific Northwest; cracks go through both AC and CTB; on fill; no groundwater; on high side of superelevated curve; March 1971.

was evident in one of the state highway sites, although this pavement was only 6 years old.

The ninth site, in Humboldt County near Eureka, California, was a special "experimental" pavement that had been reconstructed by the County Public Works Department (after two failures using standard roadbed and filter materials) on a highly permeable macadam type of base drainage layer with a coefficient of permeability between 10,000 and 30,000 ft/day.

Using the estimated time for complete drainage of the Humboldt County roadbed after saturation as a standard of 1.0, the amounts of time needed for comparable drainage of the other sites were then expressed as ratios of their times to the Humboldt County site's time. For the Humboldt County site, a drainage time of approximately 0.01 day or 15 to 20 minutes was estimated as the amount needed for excess water to drain out of the section (this was determined from the actual times recorded for water to flow from drilled holes to outlets). In contrast, the state highway sites (excluding Michigan, which is a special case) have drainage times estimated to be in the range of 5 to 20 days for *relative* times of 500 to 2000 in comparison with the Humboldt County site. This is a spectacular difference.

In developing Table 4.2 efforts were made to make reasonable allowances for losses of water from all possibilities such as $O_S$, $O_P$, and $O_R$. The FHWA report states that actual drainage times of the state sites were probably longer than estimated, because the estimates were considered to be somewhat optimistic for the levels of permeability involved. It is, therefore, felt that the drainage rates of most "conventional" pavements are probably slower than indicated by Table 4.2.

### Lateral Drainage through Shoulders

Most of the materials normally used in shoulder and base construction are fine-grained, relatively dense, and low in permeability. These materials do not provide much beneficial drainage. Figure 4.21*a* shows a standard subbase course of a major state highway in a western state during construction. Water is still standing on the base more than a day after it has rained. Figure 4.21*b* shows a typical, sandy subbase of a type used under many pavements and shoulders. The fine texture of the material that has been placed here is apparent from the smoothness of the tire marks of construction equipment.

In some cases, pavement designs have made use of very permeable untreated or treated granular materials in shoulders in the hope of improving structural section drainage of pavements. A state in the Pacific Northwest uses ¼ to 1½ in. gravel "ballast" for shoulders in some areas requiring drainage. And, a southern state has made use of a very permeable drainage blanket under the shoulders of some pavements where severe damage from excess water was occurring, as their engineers evidently felt that good shoulder drainage would reduce the problems from excess water. A design they used is shown in Fig. 4.22.

Widespread experience has shown that the use of rounded aggregates in shoulder construction often leads to severe shoving and rutting under heavily loaded vehicles. Such problems can be virtually eliminated by using crushed, highly angular materials in bases and aggregate drainage layers under shoulders (see Sec. 9.3 for discussion of the excellent performance of open-graded, angular aggregates under heavily loaded logging trucks).

To obtain an indication of the possible benefits of drainage through shoulders, the flow net in Fig. 4.23 was constructed, assuming that a granular blanket is placed in the shoulder, as shown. Assuming that the primary source of the water is from the structural section and the edge joint between the pavement and shoulder, the basic flow pattern would be as shown, and the discharge rate could be estimated with Darcy's law in the form $q = kh(n_f/n_d)$. Since in this example $h = 2$ ft, and $n_f/n_d = 0.06$, $q = 0.12k$. For several values of permeability of shoulder

**FIG. 4.21** Bases and subbases normally used can be quite impermeable. (*a*) Water standing on a "standard" type of subbase on an interstate highway in a western state one day after a shower; September 1970. (*b*) Fine-grained subbase for an interstate highway pavement under construction; Great Lakes area; October 1970.

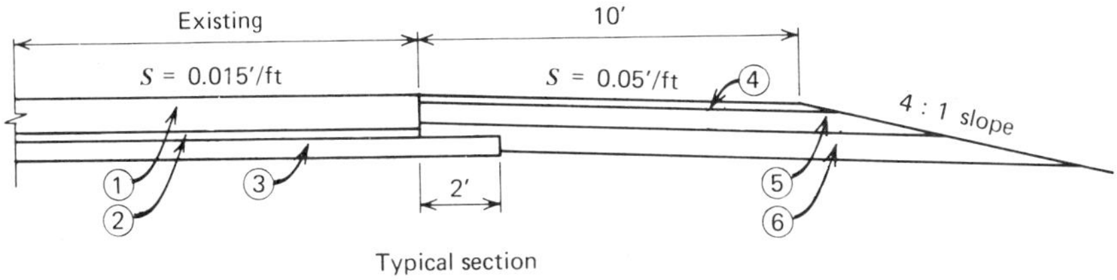


Legend

1. 10″ Existing PCC
2. 2″ Existing granular material
3. 6″ Existing soil cement
4. 2″ Type C surfacing
5. 6″ Type A drainage blanket
6. 8″ Type B drainage blanket

Specifications for type B drainage blanket

| Sieve size | Percent passing |
|---|---|
| 3/4″ | 98–100 |
| 3/8″ | 50–80 |
| No. 4 | 10–40 |
| No. 10 | 0–5 |

**FIG. 4.22** Shoulder reconstruction to improve drainage of an existing roadbed being damaged from excess water. (FHWA, 1973, *Final Report*).

drainage blanket material, the drainage rates per foot of roadway are given in Table 4.3.

Since many of the materials frequently used in constructing roadway shoulders (both highways and airfields) have coefficients of permeability in the range of 1 to 10 ft/day or less, it is evident that only minor amounts of drainage can be expected where these materials are used. Where highly permeable shoulders are constructed, there still remains the problem of water getting to the shoulders if bases and subbases under the primary pavements are low in permeability.

The author has examined pavements that had trench drains constructed along their outer edges after pumping and other water damage had started to develop and has found that relatively little help was being given to the pavements by these drains. Under these conditions, drains installed at pavement edges provide little beneficial drainage for more than a foot or two away from the drains.

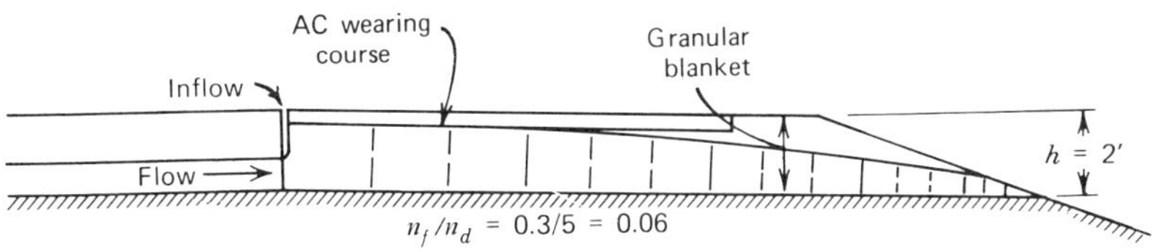


**FIG. 4.23** Flow net for seepage in pavement shoulder.

**TABLE 4.3** Possible Drainage Rates through Shoulder Drainage Blankets (for flow net in Fig. 4.23)

| Coefficient of permeability (ft/day) | Drainage rate ($ft^3$/day) |
|---|---|
| 1 | 0.1 |
| 10 | 1.2 |
| 100 | 12 |
| 1000 | 120 |

## Downward Drainage into Subgrades

If the subgrade of a pavement is highly permeable sands or gravels, it can provide significant downward drainage; however, if the subgrade is very low in permeability, downward drainage may be so slow as to be insignificant. To obtain an indication of the levels of drainage that are possible by downward seepage into the subgrades under pavements, the flow net in Fig. 4.24 was constructed. In developing the cross section for this flow net, it was assumed that the pavement surface is porous and that infiltration through the structural section will enter into the subgrade and eventually drain out of the embankment on which this road is built. When making studies such as this, it should be recognized that possible rates of drainage into subgrades depend not only on the effective permeabilities but also on the depth to the natural water table. If the water table is deep, perhaps 50 to 100 ft or more below the surface, the effective hydraulic gradient will often be in the order of 0.5 to nearly 1.0, and the rate of seepage will be considerably greater than if the water table is shallow and the effective gradient much smaller. In Fig. 4.24 the average hydraulic gradient is approximately 0.25.

From the flow net in Fig. 4.24 the potential rate of drainage from under the entire half-width of the pavement would be: $q = kh(n_f/n_d)$. For this example, $h = 14$ ft, and $n_f/n_d = 0.43$; hence, the total seepage quantity is $q = (14)(0.43)(k) = 6.0k$. Drainage rates for some common subgrade soils are given in Table 4.4 for this cross section.

Since a great many compacted subgrades have coefficients of permeability in the range of $1 \times 10^{-5}$ cm/sec (0.03 ft/day) and lower, it is evident that beneficial drainage into subgrades is often quite small.

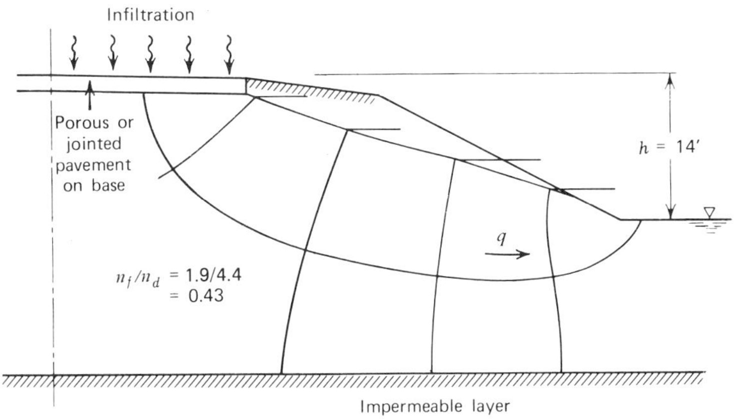


**FIG. 4.24** Flow net for subgrade seepage for pavement on an embankment.

The conditions represented in Table 4.4 are for pavements on fills with a natural water table relatively deep. Many pavements having shallower water table conditions than assumed in Fig. 4.24 and Table 4.4 will have proportionately slower subgrade drainage rates for the same permeabilities. In many locations, subgrade drainage is almost negligible, certainly not very beneficial. Figure 4.25 shows water standing on the base course of an airfield taxiway that has been dug out and is being reconstructed. There has been no rainfall for several days, so it is evident that downward drainage is very slow. As a consequence of the slow drainage of most subgrades, many pavements remain in a flooded condition long after the supply of water has been cut off.

**TABLE 4.4** Possible Drainage Rates into Subgrade for Pavement on an Embankment (cross section in Fig. 4.24)

| Subgrade permeability, $k$ | | Drainage rate |
|---|---|---|
| cm/sec | ft/day | [ft$^3$/day/lin. ft] |
| $1 \times 10^{-6}$ | 0.003 | 0.02 |
| $1 \times 10^{-5}$ | 0.03 | 0.18 |
| $1 \times 10^{-4}$ | 0.3 | 1.8 |
| $1 \times 10^{-3}$ | 3.0 | 18.0 |
| $1 \times 10^{-2}$ | 30.0 | 180.0 |

**FIG. 4.25** Airfield taxiway that has been dug out for reconstruction after failure; water is still standing several days after the last rain; September 1972. (Courtesy CERL.[1])

### Lateral Drainage of Base Courses and Base Drains

When pavements on impermeable subgrades are provided with longitudinal pipe drains along lower edges, the only significant drainage often occurs by lateral seepage to the edge drains. The geometrics of base drainage layers make them inherently difficult to drain. Little can be done about the geometrics (beyond slight increases in slope and thickness), but drainage of bases can be greatly speeded up by using extremely permeable materials in their construction. Seepage in flat-lying layers is inhibited by the smallness of the discharge area $A$, which is limited by the vertical thickness, and by small hydraulic gradients that are limited because of small vertical dimensions of these layers.

An indication of a general type of flow pattern in base drainage layers is given by the flow net in Fig. 4.26*a*. This flow net is for seepage in a permeable drainage layer under a jointed but uncracked PCC pavement. Equal quantities of water are assumed to be entering each of the three open joints; hence the number of flow channels in the flow net increases in proportion to the number of joints feeding water through the pavement. Also, it can be seen that the hydraulic gradient in the open-graded base becomes steeper in the vicinity of the collector pipe at the right. From this flow net, the minimum *required* permeability of the drainage layer is proportional to the quantity $Q/iA$.

The lower dashed line in Fig. 4.26*b* gives values of $Q/iA$ that were determined by measuring the thickness of the saturated open-graded base and the hydraulic gradient at a number of points. In general, the ratio of $Q/iA$ has a numerical value of approximately 1.3, with slight variations occurring in the right half. This line is labeled "flow net line."

Because the vertical dimensions of base drainage layers are generally quite small, it is difficult to construct flow nets for seepage in them. In fact, the vertical dimensions used in the cross section in Fig. 4.26*a* have been exaggerated to provide space to allow this flow net to be drawn. Most of the solutions in this book use the total inflow quantity and the slope of the drainage layer in estimating required coefficients of permeability with Darcy's law, recognizing that this approach involves simplifying approximations. The middle dashed line in Fig. 4.26*b* gives *relative* $Q/iA$ values obtained with Darcy's law, assuming that equal inflows enter each of the three joints. The upper dashed line in Fig. 4.26*b* gives the value of $Q/iA$ that would be used in design (using total inflow, and the slope of the base drainage layer as the hydraulic gradient). It has a numerical value for $Q/iA$ of 4.3, which is about 3 times the maximum value obtained from the flow net. In this case, the simplified method using Darcy's law has a factor of safety with respect to $k$ of about 3. In many cases, there would be little or no difference, although it is felt that the simplified method generally is slightly conservative.

Table 4.5 gives the discharge capacities of base drainage layers having a thickness of 12 in. and a slope of 0.01, as estimated with

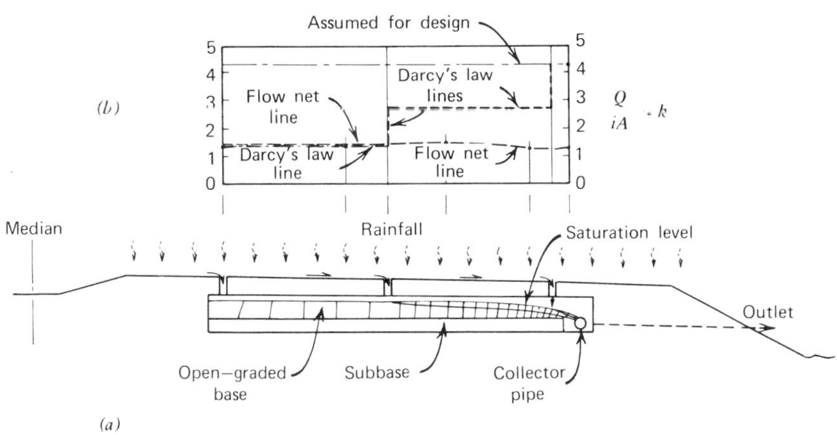


**FIG. 4.26** Flow net study of seepage in a highly permeable open-graded base under jointed PCC pavement.

**TABLE 4.5** Water-Removing Capabilities of Base Drainage Layers (thickness = 12 in., $i = 0.01$)

| Permeability of base | | Discharge capacity | |
|---|---|---|---|
| cm/sec | ft/day | [ft$^3$/day/lin. ft] | |
| $3.5 \times 10^{-6}$ | 0.01 | 0.0001 | |
| $3.5 \times 10^{-5}$ | 0.1 | 0.001 | Normal |
| $3.5 \times 10^{-4}$ | 1.0 | 0.01 | range |
| $3.5 \times 10^{-3}$ | 10.0 | 0.1 | |
| $7.0 \times 10^{-3}$ | 20.0 | 0.2 | |
| 1.0 | 3,000 | 30 | |
| 3.5 | 10,000 | 100 | |
| 7.0 | 20,000 | 200 | Recommended |
| 10.5 | 30,000 | 300 | range |
| 17.5 | 50,000 | 500 | |
| 24.5 | 70,000 | 700 | |
| 35.0 | 100,000 | 1000 | |

Darcy's law, assuming the layers are flowing full. The upper group represents construction materials of the kinds often called "permeable," for bases and drainage layers for pavements. It can be seen that their discharge capabilities are quite small. The lower group in Table 4.5 represents the macadam type of open-graded aggregate recommended in this book for drainage systems for highways and airfields. The most permeable material in this table has 10 million times the capability of the least permeable material.

Another illustration of the discharge capabilities of drainage layers (as estimated with Darcy's law) is given by Fig. 4.27, which is for layers 6 in. thick on slopes ranging between 0.005 and 0.05.

### Capabilities of Flat Layers to Discharge Infiltration

In a study of drainage of agricultural lands with tile drains at various spacings, Kirkham and deZeeuw (1952) developed the following equation for surface infiltration that can be removed by the drains:

$$q = k\left(\frac{h}{b}\right)^2 \tag{4.2}$$

In Eq. 4.2, $q$ is the uniform surface infiltration discharged to side drains, $k$ is the coefficient of permeability of the soil, $h$ is its thickness

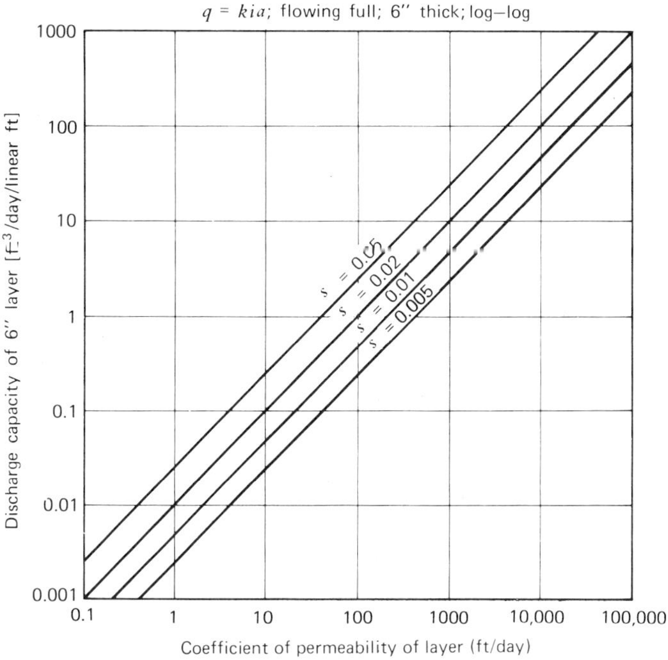


**FIG. 4.27** Discharge capabilities of bases with edge drains.

(above a horizontal impervious base layer), and $b$ is half of the distance between drains. This equation is useful in evaluating the drainage capabilities of flat bases of highways, airfields, and so forth to discharge water to the sides.

An independent derivation of Eq. 4.2 was obtained from a flow net analysis of horizontal layers on completely impermeable subgrades (Fig. 4.28) as follows:

$$Q = kh \frac{n_f}{n_d} \tag{4.3}$$

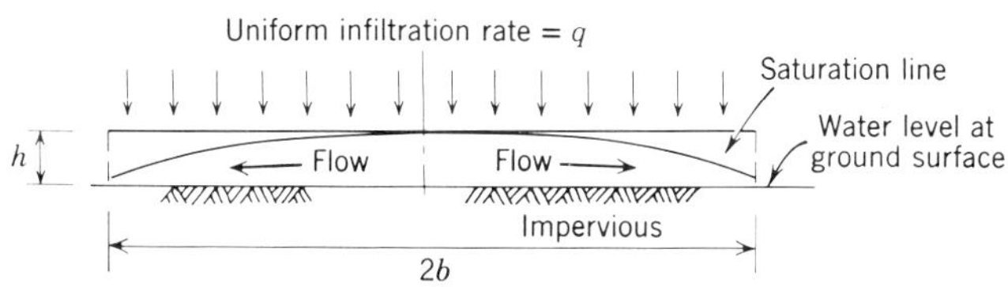


**FIG. 4.28** Cross sections of bases studied with flow nets. (*Seepage, Drainage, and Flow Nets*, John Wiley & Sons, New York, 1967, p. 350).

in which $Q$ is the total infiltration rate discharged to the sides, $k$ is the coefficient of permeability of the layer, $h$ is its thickness, and $n_f/n_d$ is the shape factor for the flow nets.

A number of flow nets for this case (see Fig. 4.29) gave shape factors that were approximately equal to the ratio of $h/b$, so Eq. 4.3 can be rewritten:

$$Q = kh\,\frac{n_f}{n_d} \qquad \text{and} \quad q = \frac{Q}{b} = k\left(\frac{n_f}{n_d}\right)\left(\frac{h}{b}\right)$$

So,

$$\frac{q}{k} = \left(\frac{n_f}{n_d}\right)\left(\frac{h}{b}\right) \tag{4.4}$$

And

$$\frac{q}{k} = \left(\frac{h}{b}\right)^2 \tag{4.5}$$

According to Eq. 4.5, it should be expected that a wide, shallow layer would be much slower draining than a narrow, relatively thick layer of the same material and permeability. In comparing the general capabilities of modern, multilane roads or wide airport pavements with

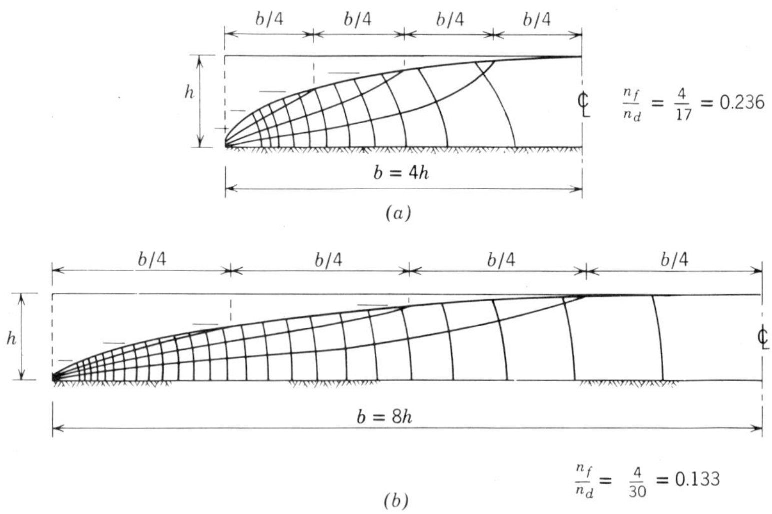


**FIG. 4.29** Typical flow nets for infiltration study of bases on impermeable foundations (cross section in Fig. 4.28). (*Seepage, Drainage, and Flow Nets,* John Wiley & Sons, New York, 1967, p. 351).

**TABLE 4.6** Relative Difficulty of Draining Pavements of Various Widths, Thickness Not Changed

| Width (ft) | $h/b$ | $(h/b)^2$ | Relative drainability |
|---|---|---|---|
| 20 | 0.05 | 0.0025 | 1 |
| 40 | 0.025 | 0.00063 | 1/4 |
| 75 | 0.013 | 0.00017 | 1/14 |
| 150 | 0.0066 | 0.000044 | 1/57 |
| 300 | 0.0033 | 0.000011 | 1/228 |

For cross section in Fig. 4.28

older, narrower pavements, it can be expected *from geometry alone* that wider pavements will be more difficult to drain than narrower ones. If only widths are changed, while thickness of bases are kept about the same, the relative difficulties of draining pavements increase with the squares of the widths (Eq. 4.2). Table 4.6 gives a comparison of the relative drainability of pavements of several widths.

Hence 40-ft-wide roadways are about 4 times more difficult to drain than 20-ft; 300-ft-wide pavements are more than 200 times more difficult to drain than 20-ft ones. This factor alone contributes significantly to the poor drainage performance of many modern wide airfield and highway pavements.

## References

American Iron and Steel Institute (1967), *Handbook of Steel Drainage and Highway Construction Products*, New York, 1967.

Armco Drainage and Metal Products, Inc. (1955), *Handbook of Drainage and Construction Products*, Middletown, Ohio, 1955.

Cedergren, H. R. (1967), *Seepage, Drainage, and Flow Nets*, John Wiley & Sons, Inc. New York, pp. 346–99.

Civil Aeronautics Administration, CAA (1956), *Airport Drainage*, U.S. Government Printing Office, Washington, D.C., November 1956.

Federal Aviation Agency, FAA (1970), *Airport Drainage*, Advisory Circular AC 150/5320-5B, Department of Transportation, Washington, D.C., July, 1970.

Federal Highway Administration, FHWA (1973), *Guidelines for the Design of Subsurface Drainage Systems for Highway Structural Sections*.

Kirkham, D. and J. W. deZeeuw (1952), "Field Measurements for Tests of Soil Drainage Theory," *Proceedings*, Soil Science of America, Vol. 16, pp. 286–93.

Lovering, W. R. and H. R. Cedergren (1962), "Structural Section Drainage," *Proceedings*, International Conference on the Structural Design of Asphalt Pavements, University of Michigan, Ann Arbor, Michigan, August 20–24, 1962, pp. 773–84.

Road Research Laboratory (1954), *Soil Mechanics for Road Engineers*, Her Majesty's Stationery Office, London, pp. 326–41.

U.S. Army Corps of Engineers (1965), *Drainage and Erosion Control—Subsurface Drainage Facilities for Airfields*, Technical Manual TM-820-2, Headquarters, Washington, D.C., August, 1965.

CHAPTER FIVE

# Criteria for Designing Subsurface Drains for Pavements

## 5.1 BASIC CONCEPTS

### Where Drains are Needed

The purpose of subsurface drains for highway and airfield pavements is to give these facilities the maximum practical protection from free water. Since all of the spots where water will eventually get into pavements cannot be predicted (open joints, shrinkage cracks, and other openings and porous areas can develop almost anywhere), *subsurface drains must be provided to protect the entire traveled widths of pavements that could be subjected to appreciable numbers of heavy wheel loads while the structural sections are exposed to the entry and accumulation of free water.* For this reason, the primary water-removing layer should extend under the full width of the traveled way of a pavement. Since no major type of pavement in widespread usage at the present time can be kept watertight for appreciable lengths of time, drainage systems are needed for both rigid and flexible types of pavements.

Many investigators have studied the types of damage to pavements that are caused by or aggravated by inadequate drainage (see Chap. 2). Edgar (1944) discusses and gives examples of the effect of subgrade moisture content, frost heaving, and moving loads on pavement condition. Morrison (1931) emphasizes that the more serious and frequent causes of pavement failure are poor subsoils or subgrades, inadequate drainage, and neglect of maintenance. The U.S. Army Corps of Engineers (1947) describes the modes of failure of various kinds of pavements and describes methods for sealing and repairing bituminous pavements and for draining pavements. The U.S. Army Corps of Engineers (1945–1946), after making a lengthy investigation of base course drainage, concluded that drainage of base courses on frost-susceptible subgrades (in cold regions) is required where the pavement and base thickness is not sufficient to prevent formation of ice lenses in the subgrade. They also concluded that, "It is possible to determine within reasonable limits the time-degree of drainage relationships of a saturated base course on an impermeable subgrade where reliable values of the coefficient of horizontal permeability are available."

Road builders from ancient times have recognized the need for controlling high groundwater, inflows from springs, and the like, and they often used drains to take water from such sources out of structural sections or to prevent it from entering, but almost never have subsurface drains been used for the removal of surface water infiltration.

After a thorough study of the occurrences and damage caused by surface water infiltration into pavements, the Federal Highway Administration (FHWA, 1973) issued a report which says,

> Subsurface drainage systems should be provided for all important highway pavement structural sections unless economic feasibility studies indicate that they are not cost-effective, precipitation quantities and frequencies are so small that drainage is not needed, or heavy wheel load impacts per day are very small.

Possible locations where subsurface drains *might not be needed* could be the following: (1) in areas in which there is no groundwater, spring inflow, and so forth, the annual rainfall is less than 8 or 10 in., and no significant amount of melting snow or ice can enter structural sections; (2) in areas in which subgrades are *very permeable* and are not subject to freezing (see "Downward Drainage Into Subgrades," Chapter 4), and natural water tables are very deep; and (3) in pavements that will be subjected to very limited numbers of heavy wheel loads over their design life, for example, on highways, less than 150 or 200 18,000-lb axle loads per day.

In view of the enormous potential benefits of good drainage in cutting maintenance and replacement costs and in extending pavement life, the burden of proof should be to determine *where drains are not needed*, rather than *where they are needed*, as has often been the practice in the past. When the highly efficient drainage systems recommended in this book are provided under the full widths of pavements, they can be designed to provide protection from unexpectedly high inflows from springs, seasonally fluctuating groundwater, or other sources of inflow *at little or no increase in cost*; hence they can virtually eliminate the risks of water damage to pavements.

### Basic Requirements of Drainage Layers

The essential components of a subsurface drainage system (see Fig. 4.1) are (a) an open-graded base drainage layer, (b) a suitable subbase or filter layer and where necessary an impervious membrane over the subgrade, (c) a longitudinal collector drain, with suitable pipes and backfill aggregate in a shallow "V" trench or a rectangular trench, (d) outlet pipes, and (e) outlet markers on suitable posts to protect outlet pipes from damage and to "flag" their locations so that none will be overlooked during periodic maintenance inspections.

To adequately protect a heavy-duty highway or airfield pavement, its subsurface drains must be capable of removing any water that enters, and they must drain fast enough to prevent harmful accumulations of free water within the structural section. This can be accomplished only with a highly permeable, macadam type of base drainage layer having a coefficient of permeability of thousands of feet a day (1.0 cm/sec and greater).

To ensure the necessary level of permeability, the base drainage layer must be constructed of clean aggregate having a relatively narrow range in particle sizes (see Open-graded bases, Fig. 5.9). Types of gradation ranges that will guarantee high permeability are $\frac{3}{8}$ in. to no. 4; $\frac{3}{4}$ in. to no. 4; $\frac{3}{4}$ in. to $\frac{1}{2}$ in., 1 in. to $\frac{1}{2}$ in.; 1 in. to no. 4; and so forth. Two criteria that will help guarantee that the open-graded materials will be of the most optimum gradations are the following.

To restrict the particle sizes of the material to a relatively narrow range, the 85 percent size (85 percent is finer) of open-graded aggregates should be less than 4 times the 15 percent size (15 percent is finer). This is expressed by the following equation:

$$D_{85} < 4D_{15} \tag{5.1}$$

To ensure a high level of cleanness, the 2 percent size (2 percent is finer) of an open-graded layer should be at least 0.1 in. in diameter. This is represented by the following equation:

$$D_2 \geqq 0.1 \text{ in.} \tag{5.2}$$

It should be emphasized that the coefficient of permeability of a drainage material should always be verified by tests made on control samples taken from the work after the material has been placed and compacted. A rough guide for estimating the required degree of coarseness of materials for open-graded base drainage layers is given in Fig. 5.1. The curves in Fig. 5.1, which relate coefficient of permeability with the 15 percent size ($D_{15}$), should be used only for rough approxi-

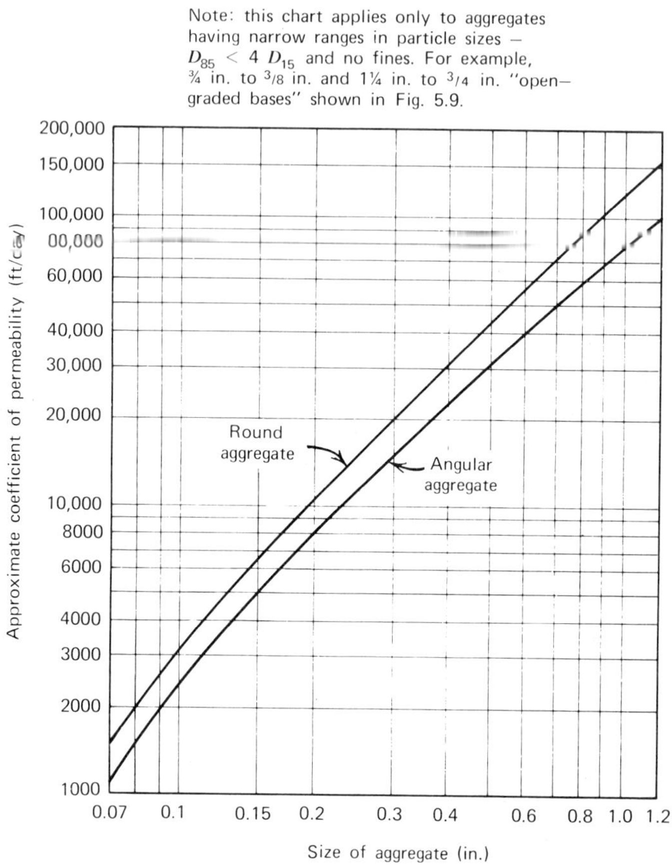


**FIG. 5.1** Rough guide for estimating coefficient of permeability of narrow size-ranged aggregates with no fines.

mations, because the permeability of an individual material can vary appreciably, depending on the types of minerals and the shapes of the particles as well as on compaction conditions and density.

When a pavement is designed with a highly efficient subsurface drainage system, it is less vulnerable to many of the incidental or only slightly related highway and airfield regimens such as general planning, general design, safety, landscaping, adjacent structures, and general maintenance practices; the drainage system will usually require only minimal attention and service. Nevertheless, the basic approach to the design of highway and airfield pavements that will not be damaged by excess water should be: (1) use good geometrics in designing pavements, shoulders, and adjacent features that will aid surface drainage, rather than impede it, whenever it is practical to do so; (2) prepare plans and specifications for pavement-drainage systems and related features that will aid in obtaining high quality workmanship, and (3) set up maintenance programs geared to the protection of the pavements, the drains, and their outlets.

The heart of a drainage system is the layer of highly permeable open-graded aggregate with pore spaces large enough to quickly convey water to longitudinal collector pipes along its lower edges and then to outlet pipes which remove it by gravity flow. In very limited cases it may be feasible to "daylight" the drainage layer. Wherever subgrades are erodible soils or soft, highly weathered rocky formations, an appropriate base or subbase must be placed between the coarse drainage layer and the subgrade to serve as a filter to keep soil fines out of the open-graded layer (in some cases, special filters or membranes may be required).

Barber and Sawyer (1952) pointed out two decades ago that, "The small gradients available for lateral drainage of base courses prevent rapid drainage of dense-graded materials on impervious subgrades." And, "Open-graded materials will drain more rapidly; however, they must be protected from intrusion of fine subgrade soil."

Rather than resorting to some "standard" designs for these systems, each case should be evaluated on its own needs, using the rational seepage analysis methods to estimate the quantities of water that may need to be removed and to analyze other drainage requirements of the water-removing layer.

To make sure that harmful accumulations of excess water in a structural section and its drainage system cannot occur, the discharge capabilities of all important components should *increase* in the direction of flow, *never decrease*. In designing a drainage system, a check should always be made to be sure that no restrictions of outflow capabilities are built into the system. In the operation and maintenance of a pavement-drainage system, no blockages or restrictions in collector pipes, outlet pipes, or other drainage components should be allowed. Application of a principle referred to in Chapter 3 as a *condition of continuity* requires that the basic adequacy of all components of a drainage system be consciously checked by designers.

If damage from exposure to free water is to be kept to a minimum, the transmissibility of a drainage layer should always be adequate to remove all of the estimated inflows less any beneficial outflows (Sec. 5.2). The length of time needed for water to flow completely through the drainage layer should be small enough to prevent damage—especially important in cold regions where freezing can create problems (Sec. 5.3). Also, the time for saturation mounds to drain away after inflows stop should not be excessive (Sec. 5.4). In most cases if a drain has sufficient permeability to allow water to flow out in an hour or so, saturation mounds will dissipate quite rapidly after the inflows stop.

## 5.2 QUANTITIES OF WATER THAT CAN BE REMOVED

One of the basic requirements of a subsurface drain is that it must be capable of removing all the water that reaches it, otherwise excess water can accumulate under or within the structural section, and damage to the pavement may occur. As described in Chapter 4 under "Lateral Drainage of Base Courses and Base Drains," the discharge capabilities of a drainage layer can be estimated with Darcy's law $Q = kiA$, in which $Q$ is the quantity of water that can be removed from a drainage layer having a coefficient of permeability $k$, a hydraulic gradient $i$ (approximately equal to the slope of the layer), and a cross-sectional area $A$, normal to the direction of flow. Usually $Q$ is calculated per linear foot of length of road or width of drainage layer. Then the area $A$ is equal to its thickness $t$ times 1.0 ft, which is numerically equal to its thickness. Thus a thickness of 0.5 ft represents an area of 0.5 ft.$^2$

In order for a base drainage layer to be able to remove an estimated quantity $Q$, it must have a *transmissibility* $kA$ which can be estimated from Darcy's law in the form:

$$kA = \frac{Q}{i} \tag{5.3}$$

When seepage quantities are being determined per linear foot, the transmissibility $kA$ is also equal to $kt$, in which $t$, is the clear, unobstructed thickness of a drainage layer. Figure 5.2 is a chart of transmissibilities in ft$^3$/day for drainage layers of a wide range in coefficient of permeability $k$ and a range of clear thicknesses $t$. By referring to Fig. 5.2, it can be seen that for a given transmissibility, the required thickness of a layer diminishes in proportion to the amount the permeability increases. Thus a transmissibility of 10,000 ft$^3$/day can be obtained by using a 1.0-ft thick layer with a coefficient of permeability of 10,000 ft/day, a 0.5-ft layer with a coefficient of permeability of 20, 000 ft/day, a 0.25-ft layer with a permeability of 40,000 ft/day, and so forth. It is thus apparent that the greatest theoretical economy can be achieved by using materials with the highest practicable coefficients of

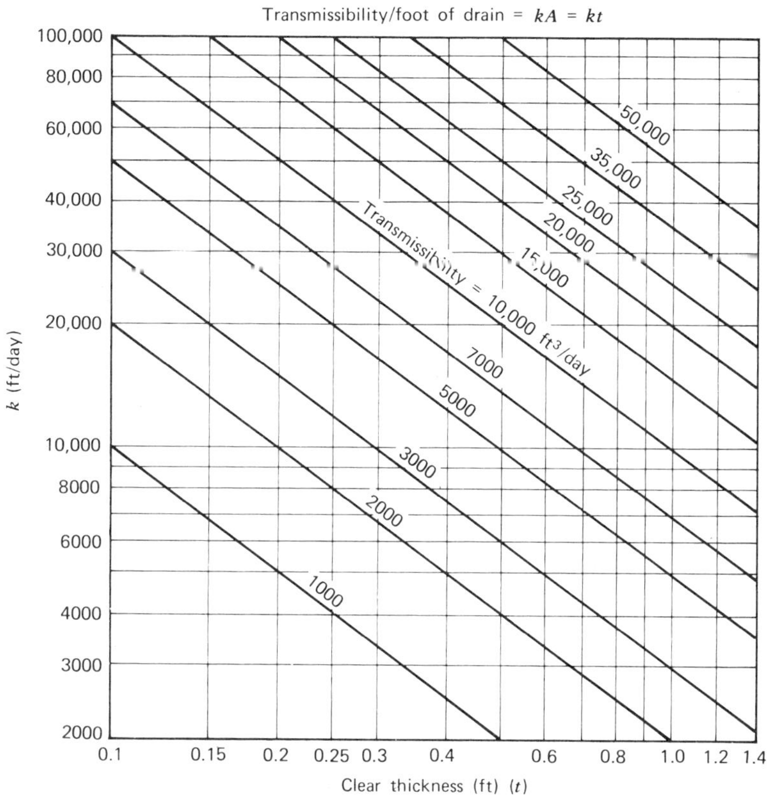


**FIG. 5.2** Transmissibility of drainage layers—cu ft/day/foot.

permeability. As pointed out in other places in this chapter and in Chapter 6, construction feasibility, design considerations, and other factors often influence selection of the most economical and practical combination of permeability and thickness for individual project situations.

Other useful charts can be developed by using Darcy's law. Thus Fig. 5.3 is an illustrative chart giving total inflows removed per foot of pavement (including surface infiltration and any other inflows converted to equivalent unit infiltration) for pavements of various widths with drainage layers of a wide range in permeability. The base drainage layers are assumed to have an unobstructed thickness of 0.5 ft, and a slope of 0.01 is assumed equal to the average hydraulic gradient in a drainage layer. From Fig. 5.3 it is apparent that some of the "standard" types of base courses or drainage layers in widespread usage in pavement construction have permeabilities that permit the re-

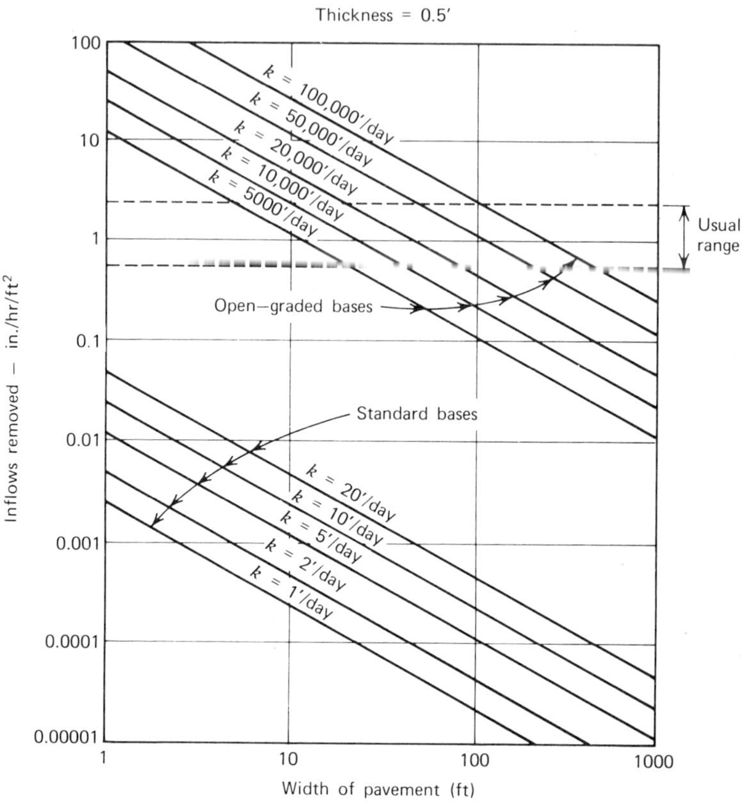


**FIG. 5.3** Capabilities of bases with edge drains to remove infiltration. Lower lines are for standard types of bases; upper lines are for open-graded bases (s = 0.01).

moval of only minute quantities of water—actually less than a light drizzle—and therefore these materials have no significant drainage value (see lower sloping lines). In contrast, the macadam types of open-graded materials recommended in this book for pavement drainage (represented by the upper sloping lines) can remove very significant rates of inflows from multilane highways and wide airport runways, taxiways, parking aprons, and so forth. In fact, there is probably no practical drainage situation that cannot be handled by materials of this kind in well-designed systems.

Referring again to Fig. 5.3, it is seen that a 6-in.-thick layer of macadam aggregate with a coefficient of permeability of 50,000 ft/day (about 17 cm/sec) can remove a total inflow rate of 1 in./hour from a pavement about 100 ft wide.

Another chart, Fig. 5.4, can be used for directly determining the required thickness and permeability of a subsurface drainage layer (open-graded) that will remove a specified design infiltration rate $I$. The recently issued FHWA *Guidelines* (FHWA, 1973) recommends that drainage systems for highways have discharge capacities based on a design precipitation rate that is equal to the 1 hour/1 year frequency precipitation rate, as determined from weather maps or weather records for a given climatic area, and that a factor of $\frac{1}{3}$ to $\frac{2}{3}$ be applied to determine the design infiltration rate $I$. The magnitude of the factor used depends on the designer's estimate of the proportion of the 1 hour/1 year frequency precipitation rate that he considers will enter a given payment after traffic and weathering.

In Fig. 5.4 the ordinates are the geometric factor $W/s$, and the abscissae are coefficients of transmissibility $C$, in inch-feet/day. Here, $W$

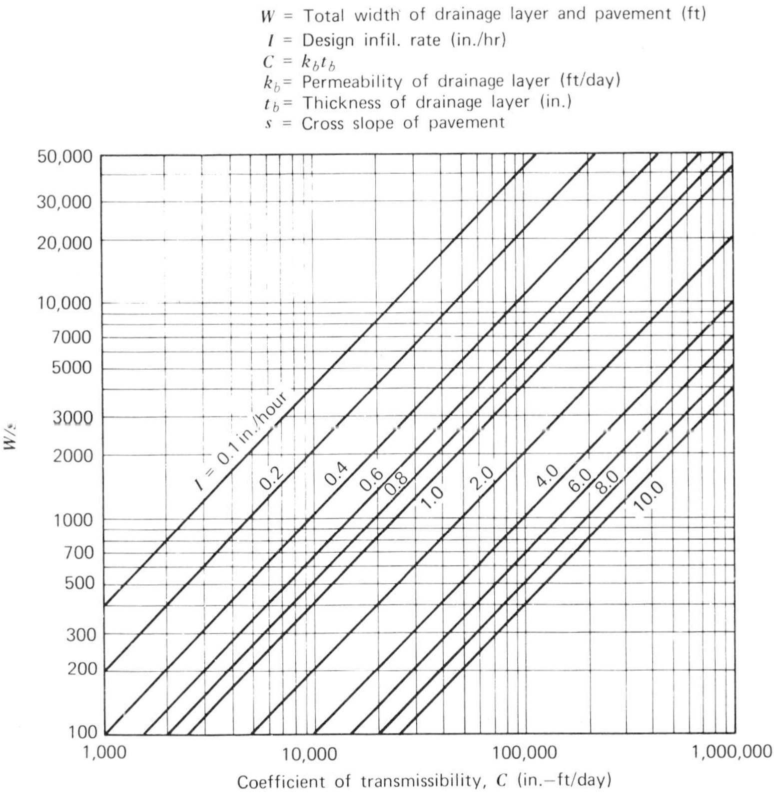


**FIG. 5.4** Coefficient of transmissibility versus $W/s$ ratio (from FHWA, 1973).

is the width of a pavement, and $s$ is its cross slope. *The ratio $W/s$ is also equal to the seepage distance divided by the slope in the direction of seepage for pavements having a longitudinal slope as well as a cross slope* (see the typical drainage analysis calculations in Sec. 6.3). The use of Fig. 5.4 may be illustrated as follows: Assume it has been determined that a subsurface drainage system should contain an open-graded water-removing layer with an unobstructed thickness of 8 in. If the pavement width $W$ is 200 ft, and its cross slope $s$ is 0.01, the ratio of $W/s = 200/0.01 = 20{,}000$. Then, if the design infiltration rate $I$ is 0.5 in./hour, the required coefficient of transmissibility $C$ is read from the chart to be 250,000. Since $C = k_b t_b$, the *required* minimum value for $k_b = C/t_b = 250{,}000/8 = 31{,}000$ ft/day (about 10 cm/sec).

In designing a drain, the actual thickness of the drainage layer should be 1 in. greater than the theoretical thickness determined from Darcy's law, to allow for some contamination at upper or lower surfaces.

Determination of the required transmissibility and suitable thickness and permeability of a drainage layer can, of course, be made by direct, basic calculations with Darcy's law, using cross-sectional areas, slopes, and seepage distances along flow paths. This method is used, in addition to charts, in examples given in Chapter 6.

## 5.3 TIME REQUIRED FOR WATER TO FLOW THROUGH DRAINAGE SYSTEMS

Another factor that should be checked in designing a subsurface drainage system is the maximum amount of time that any particles of free water that enter can stay in the system before reaching an outlet pipe. This determination is particularly important in cold regions, where frost heaving and other problems from freezing of water in the structural section of an airfield or highway pavement are important. Figure 5.5*a* shows the shoulder of an interstate highway that has been seriously damaged from frost action, and Fig. 5.5*b* shows the shoulder of an airfield taxiway in a cold region where slow drainage allowed frost damage to occur. In all such cases, no significant amount of water that enters should remain long enough to freeze during sudden drops in air temperature. This is especially important during periods of fluctuating temperatures—nighttime freezing and daytime warming with showers or melting snow and ice to supply inflows.

As explained in Chapter 3 under "Time Required for Water to Flow Through Pavement-Drainage Systems," travel times for drops of water to flow through a base drainage layer can be estimated with Darcy's

**FIG. 5.5** Rapid drainage is needed to prevent frost damage. (*a*) Severely deteriorated AC shoulder of an interstate highway in northern state; frost heaving was caused by trapped water; September 1972. (*b*) Deteriorated shoulder of an airfield taxiway; frost damage increased by liberal water supply and slow drainage; Great Lakes area; September 1972. (Courtesy CERL.[1])

law in the form, $t = S/v_s$ (Eq. 3.10), with $t$ being the travel time, $S$ the maximum seepage distance in the base drainage layer, and $v_s$ the seepage velocity ($v_s = ki/n_e$).

For any assumed travel time, the required coefficient of permeability $k$ of an individual drainage layer can be obtained by direct calculations with Darcy's law arranged in the form, $k = v_s n_e/i$. However, Fig. 5.6 is a chart that facilitates this check for highway pavements, as common geometrics are represented. Figure 5.6 has families of curves for one-lane, two-lane, three-lane, and four-lane pavements (all sloping in one direction) with drainage layers from 3 to 9 in. thick over a wide range of longitudinal grades. It is for highway pavements having a cross slope of 2 percent ($s = 0.02$), an effective porosity of 0.2 ($n_e = 0.2$), and for an allowable maximum drainage time of 1 hour.

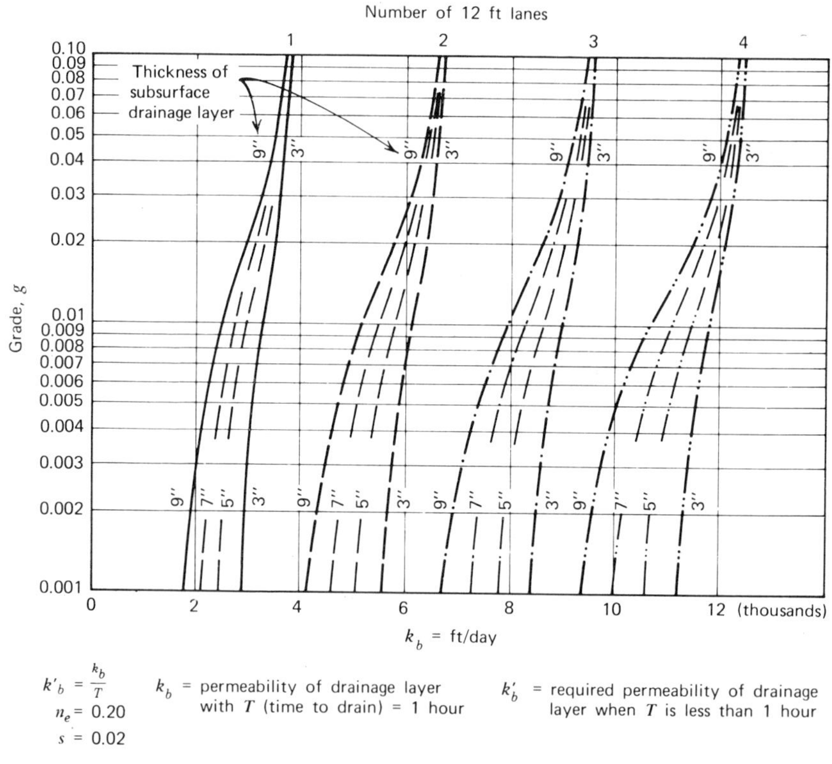


**FIG. 5.6** Permeability required in order to drain subsurface drainage layer in 1 hour or less (FHWA, 1973).

The use of Fig. 5.6 is illustrated as follows: Assume that a three-lane pavement with a longitudinal grade of 1 percent ($g = 0.01$) has a base drainage layer 6 in. thick that needs to have a minimum coefficient of permeability of 20,000 ft/day to remove the design infiltration rate. Will this layer (with $k = 20{,}000$ ft/day) remove water sufficiently fast to satisfy the requirement that the maximum travel time will not exceed 1 hour? Entering the appropriate family of curves in Fig. 5.6 (three lanes, 6-in. thickness, and grade $= 0.01$), a $k$ value of 8,500 ft/day is read at the bottom of the chart. This is less than 20,000 ft/day, so the design is more than satisfactory according to this criterion.

If a shorter travel time is required (as for cold regions), the minimum allowable permeability is obtained from the formula:

$$k'_b = \frac{k_b \text{ (from chart)}}{T} \tag{5.4}$$

In Eq. 5.4, $k_b$ is the permeability value obtained from Fig. 5.6 for a maximum travel time of 1.0 hour, and $k'_b$ is the permeability needed for the modified travel time $T$. Thus in the preceding example if $T$ is required to be 0.5 hour, $k'_b = 8{,}500/0.5 = 17{,}000$ ft/day.

Figure 5.7 is a chart which allows determination of the required permeabilities of drainage layers for airfield pavements that limit the travel time to a maximum of 2 hours. Curves are given in Fig. 5.7 for pavements with cross slopes of 0.005, 0.01, 0.015, and 0.02 (with flat longitudinal grades), and base drainage layers with an effective porosity of 0.3 and thicknesses of 0.5 and 1.0 ft. It is assumed that longitudinal pipe drains are located under lower, outer edges, with free gravity outlets.

According to the criterion of a 2-hour travel time as in Fig. 5.7, it can be seen that taxiways with half widths of 37.5 feet should contain drainage materials with coefficients of permeability in the order of 5,000 to 10,000 ft/day (1.7 to 3.5 cm/sec), but runways or parking aprons with drainage distances of 150 to 200 ft would need base drainage materials with coefficients of permeability in the range of 30, 000 to 100,000 ft/day (10 to 35 cm/sec), with actual values depending on the cross slope and the thickness of a given layer.

If a subsurface drain is to have the capability to limit the travel time of drops of water to 1 hour or less (for highways) and 2 hours or less (for thicker, wider airfield pavements), the analysis presented here indicates that the coefficient of permeability of the drainage layer must be thousands of feet a day. This requires the use of "pea gravel" and *coarser* one-sized, or narrow range macadam type, aggregate. The re-

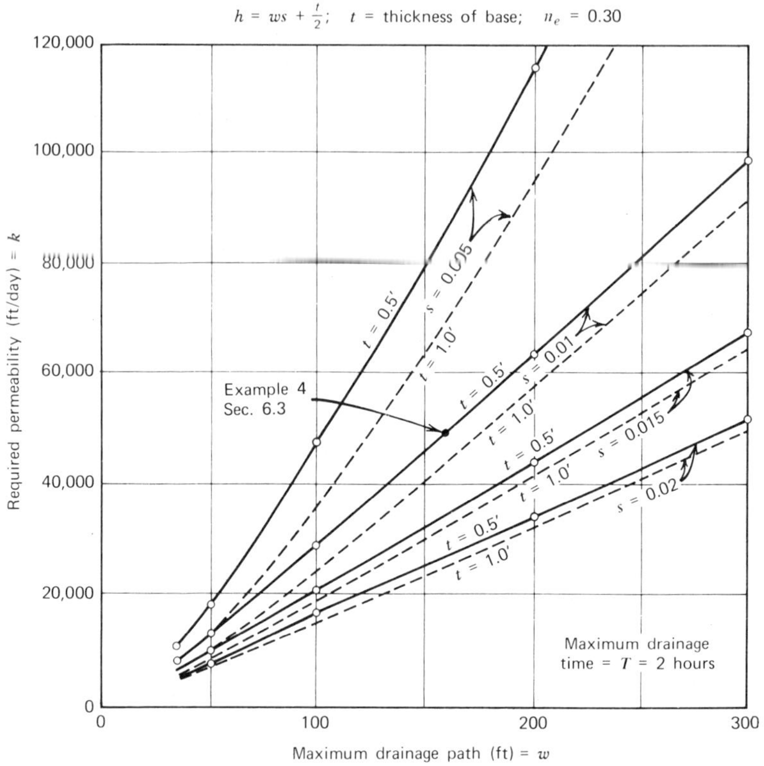


**FIG. 5.7** Minimum permeability required in order to drain base in 2 hours or less. (Courtesy CERL.[1])

quired permeability can be determined by seepage velocity calculations with Darcy's law, or from charts such as those presented, which were obtained by such calculations. The coarse materials needed for rapid drainage are very nonsusceptible to freezing problems. Obviously, the "standard" well-graded blends of sand and gravel widely used for roadbed construction (as typified by the sloping lines in the lower part of Fig. 5.3) would be very ineffective in protecting roadbeds of highways or airfields from freezing problems caused by free water being trapped in structural sections. Severe pavement distress often occurs in pavements in cold regions when they are constructed with low-permeability, frost-susceptible bases. This has led to the practice of using nonfrost-susceptible granular materials to try to prevent freezing problems. If these materials are permitted to remain full of water just prior to low temperature weather, freezing problems may still develop (Fig. 5.5, for example).

## 5.4 TIME FOR FALL OF SATURATION AFTER STOP OF INFLOWS

When the highly permeable, open-graded macadam types of base drainage layers are used under pavements, drainage is usually very rapid, and any saturation that builds up is very temporary, limited in most cases to the periods of heavy inflows. If these layers are sufficiently permeable to drain very rapidly, the total annual exposure to excess water will generally be very limited. If materials with much lower permeabilities are used, however, saturation levels may rise very high and fall very slowly after inflows have ceased. To add up the total number of hours per year that free water can remain in a structural section (from surface infiltration) it is necessary to estimate (1) the number of hours of saturating rainfall (and other surface water supplies such as from melting snow or ice) and (2) the time lag of the fall of saturation in the structural section after the supply has been cut off. The number of hours per year that local rainfall rates will exceed any given amount can be estimated from rainfall records. The rainfall rate that can keep a given pavement filled with water will vary with the width of the pavement, its slope, its surface runoff characteristics, the components of the structural section including their permeabilities, the permeability of the subgrade, and the depth to the water table. Designers should estimate values they feel are reasonable. Many of the conventional slow-draining structural sections on low permeability subgrades cannot even dispose of steady infiltration rates as small as 0.001 to 0.01 in./hour, but highly permeable base drainage layers can discharge rates in the order of 0.5 to 10 in./hour.

To estimate the amount of time that free water can stay in a structural section after each "saturating" rainfall event, the time lag of the fall of saturation is needed. Such calculations can be made with the aid of transient flow nets or with Darcy's law with step-by-step procedures, using lowering saturation mounds. A useful formula, developed from model tests and seepage theory, for calculating the rate of fall of saturation in airfield pavement base courses with edge drains is given in a U.S. Army Corps of Engineers (1965) Technical Manual.

That manual gives the following formula:

$$t = \frac{n_e D^2}{2880\, k H_o} \tag{5.5}$$

In Eq. 5.5 $t$ is the time for 50 percent drainage of a sloping base course with an edge drain along its lower edge; $n_e$ is the effective porosity of the base, $D$ is its sloping width, $k$ is its coefficient of permeability in feet/minute, and $H_o = H + sD$, where $H$ is the thickness of the base in feet, and $s$ is its cross slope. When coefficients of permeability are

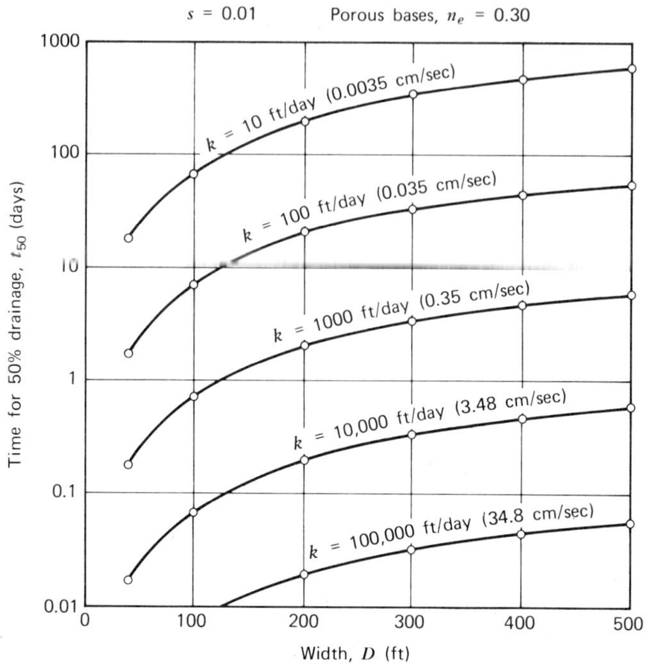


**FIG. 5.8** Permeability versus time for 50% drainage of bases with edge drains. (Courtesy CERL.[1])

expressed in ft/day, Eq. 5.5 becomes:

$$t = \frac{n_e D^2}{2\,kH_o} \tag{5.6}$$

Figure 5.8 was prepared from Eq. 5.6 to relate the amount of time needed for 50 percent drainage of pavements having bases 12 in. thick on slopes of 1 percent ($s = 0.01$) and an effective porosity of 30 percent ($n_e = 0.3$). By examining Fig 5.8, it is seen that if a criterion of 50 percent base drainage in 0.1 day (2.4 hours) is assumed, the permeability would need to be approximately 2000 ft/day for taxiways with half-widths of 40 feet. For the wider dimensions of most airfield pavements, a criterion of 0.1 day for 50 percent drainage would require that $k$ values of 10,000 to 80,000 ft/day be obtained.

## 5.5 FILTER PROTECTION OF OPEN-GRADED DRAINAGE MATERIALS

Where coarse, open-graded macadam types of bases are used for pavement drainage on erodible subgrades, these layers must be protected from the intrusion of fine material that can clog pore spaces and reduce permeability. The "inverted filter" invented by K. Terzaghi for the prevention of heave and piping failures of dams, sheet pile walls, and so forth (Terzaghi, 1922, 1929), utilizes a coarse water-removing layer placed over an appropriate "filter" that keeps fine-grained soils from working up through the coarse material. This principle has been widely used in the design of earth dams and other water-retaining structures for the past 30 or 40 years. Experimental work by Bertram (1940), the U.S. Bureau of Reclamation (Karpoff, 1955), the U.S. Corps of Engineers (1941), and others has produced practical criteria for the design of filters for dams, pavements, and other structures.

The work of Bertram and the corps of engineers led to the following criterion for filter design:

$$\frac{D_{15} \text{ (of filter)}}{5} \leqq D_{85} \text{ (of soil)} \tag{5.7}$$

In Eq. 5.7, $D_{15}$ is the 15 percent size (15% is finer than this size) of the filter layer that is protecting a soil with an 85 percent size of $D_{85}$.

To assure reasonable uniformity of grading and to prevent the use of gap-graded materials, the following additional criterion has also been used:

$$\frac{D_{50} \text{ (of filter)}}{25} \leqq D_{50} \text{ (of soil)} \tag{5.8}$$

Use of these criteria assures a high level of filter protection, because it is virtually impossible for soil fines to work into or through uniformly mixed filter materials conforming to Eqs. 5.7 and 5.8. A coarse, open-graded drainage layer should never be placed in direct contact with nonstabilized sand, silt, clay, or any other erodible subgrade soil that can be washed into the bottom or sides of the layer by seepage or be forced into it under the pressures or impacts of traffic. Nor should any aggregate or paving material be in direct contact with the top of an open-graded layer if it can penetrate into the layer. When open-graded drainage aggregates are placed over or under many of the normally used bases or subbases, the foregoing filter criteria (Eqs. 5.7 and 5.8) will often be satisfied. Nevertheless, a check of the compatibility of adjacent materials should always be made, and their gradations should be modified or a thin choker or filter layer placed between them and the open-graded material whenever necessary to ensure that no migration or intrusion of any fines or foreign matter can take place *from any direction.*

Figure 5.9 gives examples of gradations of bases or subbases that are compatible with two sizes of open-graded drainage materials.

Figure 5.9*a* has an open-graded drainage layer with a size range of $\frac{3}{4}$ to $\frac{3}{8}$ in., and a compatible filter layer that could range in sizes from a maxium of around $1\frac{1}{4}$ in. down to about the no. 50 sieve, or from a maximum size of about $\frac{3}{4}$ in. down to the no. 100 sieve. If no groundwater seepage can occur, larger amounts of fines could be permitted in the base or subbase (filter layer) as indicated.

In Fig. 5.9*b* the open-graded base drainage layer ranges from $1\frac{1}{4}$ to $\frac{3}{4}$ in. in size. Slightly coarser base or subbase materials (filter ma-

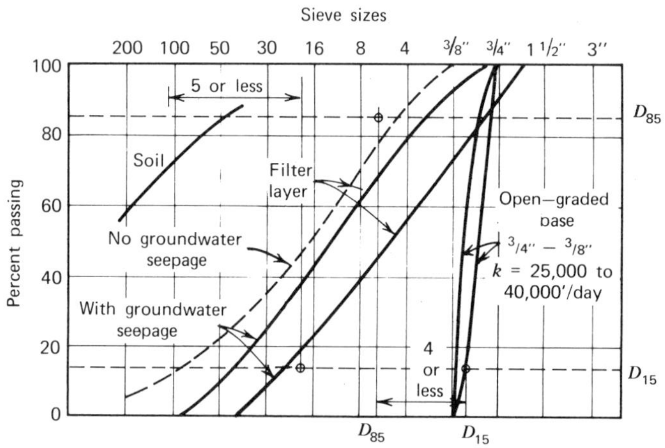


(*a*) Open–graded base constructed with 3/4" – 3/8" aggregate.

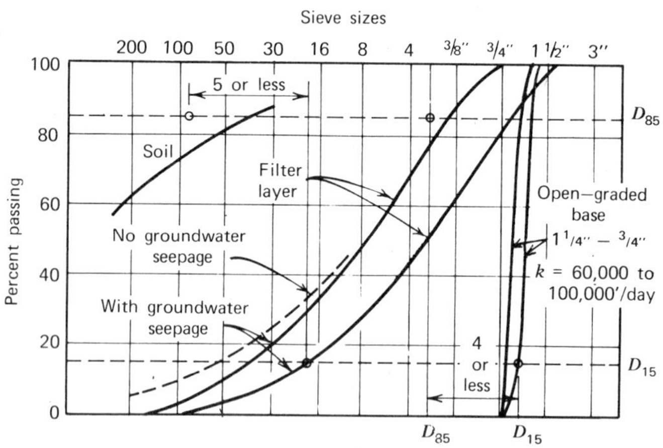


(*b*) Open–graded base constructed with 1 1/4" – 3/4" aggregate.

**FIG. 5.9** Examples of gradations of open-graded bases and their protective filters.

terials) could be used with an open-graded base of this coarseness t… with the finer open-graded base material.

If large groundwater flows can be expected, the capabilities of the base or filter layer to accept flows of water should be verified by calculations with Darcy's law *for flow through the thin dimensions of such a layer.* In estimating the capabilities of bases and subbases (or filter layers) to accept groundwater, flows from springs, and so forth, the flow per square foot can be estimated from the expression,

$$q = ki \tag{5.9}$$

For flow through the thin dimensions of these layers it is usually reasonable to allow a hydraulic gradient of about 0.5. Then the amount of water that can enter each square foot of open-graded drainage layer is about equal to $0.5k$, where $k$ is the in-place coefficient of permeability of the filter or subbase separating the open-graded layer from the water source.

If negligible or zero groundwater flow can occur, it is not necessary to verify the inflow capability across a base, subbase, or filter by the method just described, and the gradations of these layers need verification only with respect to their capabilities to serve as filters to protect open-graded layers from clogging.

Various types of plastic filter cloths can be used as filters to protect open-graded drainage layers and drain trench backfill. Extensive testing of such materials has been done by the Waterways Experiment Station, U.S. Corps of Engineers (Calhoun, et al., 1971; Calhoun, 1972). At any location where a filter cloth is proposed, the suitability of any proposed material should have the approval of a recognized testing laboratory. The choice between aggregates or plastic cloths for filter protection is largely a matter of construction feasibility and economics.

When perforated or slotted pipes are placed in drainage systems for the collection or removal of water, there should be no unplugged ends, and the drainage materials in contact with the pipes must be coarse enough so that no appreciable amount of this material can enter into the pipes. The criteria used by the U.S. Army Corps of Engineers (1965) for pipes with slots and circular holes are as follows:

For slots:

$$\frac{85\% \text{ size of filter material}}{\text{slot width}} > 1.2 \tag{5.10}$$

For circular holes:

$$\frac{85\% \text{ size of filter material}}{\text{hole diameter}} > 1.0 \tag{5.11}$$

Other agencies use these criteria or some that may vary slightly from these standards. For example, the U.S. Bureau of Reclamation (1965) specifies that the 85 percent size ($D_{85}$) of the aggregate nearest the pipe should be 2 or more times the size of the maximum opening in a drain pipe.

## 5.6 PROTECTION OF SENSITIVE SUBGRADES

In areas where highway or airfield pavements are constructed over highly expansive soils, expansive shales, or other similar geological formations, noncohesive, uniformly graded materials such as loess or other sensitive materials, it may be necessary to provide an impermeable seal under the structural section (see Fig. 4.1) to prevent surface water from soaking downward into these materials. A number of types of seals such as catalytically blown asphalt and plastic sheeting can be used to keep water out of sensitive subgrades. In some instances lime or cement has been successfuly used to stabilize such materials.

The soils investigation for important pavements should include sufficient tests to identify any soil or rock formations that would be susceptible either to swelling, collapsing, or other detrimental changes caused by water infiltration; the soils report should include recommendations for any special treatments required (impermeable membranes, stabilization, etc.). Unless adequate measures are taken, uneven swelling can cause warping of concrete slabs, cracking, or uneven humps that can be both hazardous and discomforting to the users of the pavements.

## 5.7 TRENCH BACKFILL PERMEABILITY REQUIREMENTS

In designing subsurface drainage systems it is important that any water that enters any part of a system be unrestricted in its flow toward a discharge exit. A basic concept, called a "condition of continuity" is described in Chapter 3. Equation 3.8 is an expression of that concept. In essence it says that in progressing along possible paths of flow in drainage systems, the water removing capabilities should get greater, never smaller, in the direction of flow. This is particularly important with respect to pipe drains and the backfill surrounding them. If low permeability backfill is used, or dirt or other matter is allowed to contaminate the top of the backfill, the flows to the pipes can be greatly restricted, if not virtually cut off.

In cases where drain pipes can be placed in shallow "V" trenches under the outer edges of open-graded base drainage layers (see Fig.

6.7*a, b,* and *c*), excellent flow to the pipes is assured. High water tables, other groundwater conditions, or deep freezing, however, will often require the use of deeper trench drains. If other kinds of materials are selected for backfilling the trench (for economic or other reasons) it is necessary to determine that the backfill material is permeable enough to assure free flow of water to the pipes. Wherever sufficiently permeable backfill material is too coarse to be placed next to erodible soils, suitable plastic filter cloths will be needed on the sides and bottom of the trenches (see Sec. 5.5, "Filter Protection of Open-Graded Drainage Materials").

Wherever collector pipes are placed in trenches (see Fig. 6.7*d* and *e*), a check of backfill permeability requirements can be made with Darcy's law, using appropriate dimensions and hydraulic gradients together with the estimated discharge rates, to determine the minimum required permeability of the trench backfill. In Darcy's law, $k$ is represented by $k_t$ (the permeability of the trench backfill), $A$ is equal to the trench width in feet $B$ times 1.0 ft, and the inflow rate in cubic feet per day per linear foot is $Q$. Since flow is essentially vertically downward, the hydraulic gradient is unity or nearly unity. In such calculations a liberal factor of safety should be used to allow for possible imperfections in entrance conditions into the trench.

Figure 5.10 is a chart that can be used for directly reading the required combinations of $B$ and $k_t$ that will assure sufficient capability for downward flow in collector trenches. This figure is from the FHWA *Guidelines* (FHWA, 1973) redrawn to a different scale and modified slightly.

To use Fig. 5.10 enter the chart along the abscissa to the length of pavement strip contributing flow to a drain, extend vertically to the line corresponding to the design infiltration rate (total combined inflows from surface and groundwater etc., expressed in equivalent infiltration rates) and project horizontally to the left side of the chart to read $Bk_t$, the product of trench width $B$ and trench backfill permeability $k_t$. Assume, for example, that a 40-ft width of pavement is contributing flow to a drain, and the design infiltration rate is $I = 2$ in./hour; then a value of $Bk_t$ of approximately 3200 ft$^2$/day is read from the chart. If the trench width is 1.5 ft, the required value of $k_t$ is $3200/1.5 = 2100$ ft/day. This would require the use of clean, washed pea gravel, stone chips, or equivalent. If the soil in which the trench is to be excavated is an erodible type, it will be necessary to line the sides and bottom with an approved, satisfactory type of filter cloth (see Fig. 6.8) because pea gravel, stone chips, or coarser material cannot be placed against such soils without filter protection and it is generally not practical to line such trenches with fine-filter aggregate.

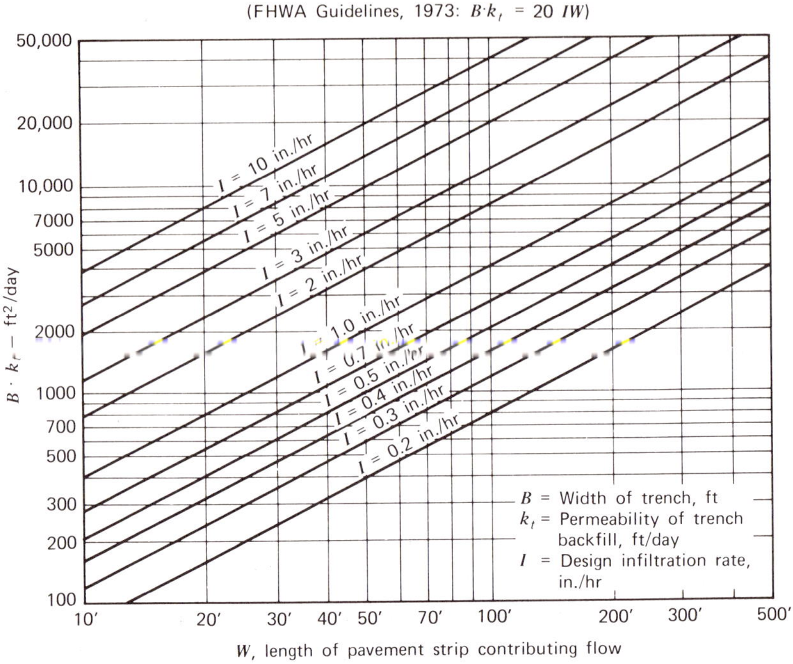


**FIG. 5.10** Chart to determine either required collector pipe trench permeability or collector pipe trench width.

**FIG. 5.11** Interstate highway in Pacific Northwest constructed on very permeable base, but with no drainage outlets; "bathtub" condition has caused severe damage; October 1970.

## 5.8 COLLECTOR PIPE DIAMETER AND SPACING

Suitable collector pipes and outlet pipes are vital parts of essentially every pavement subsurface drainage system, because without an adequate outlet system no drainage facility can function properly. Figure 5.11 shows a section of an interstate highway in the Pacific Northwest that was constructed over a very coarse gravel subbase of very high permeability ($k$ at least 50,000 ft/day), but with *no outlets of any kind.* As a consequence, water collected under lower edges and at sag vertical curves, causing severe damage to the pavement. Suitable slotted or perforated pipes should be placed along lower outer edges and at other locations as required to prevent the entrapment of water in drainage systems. Outlets should be located wherever needed to ensure the free flow of water to safe exits.

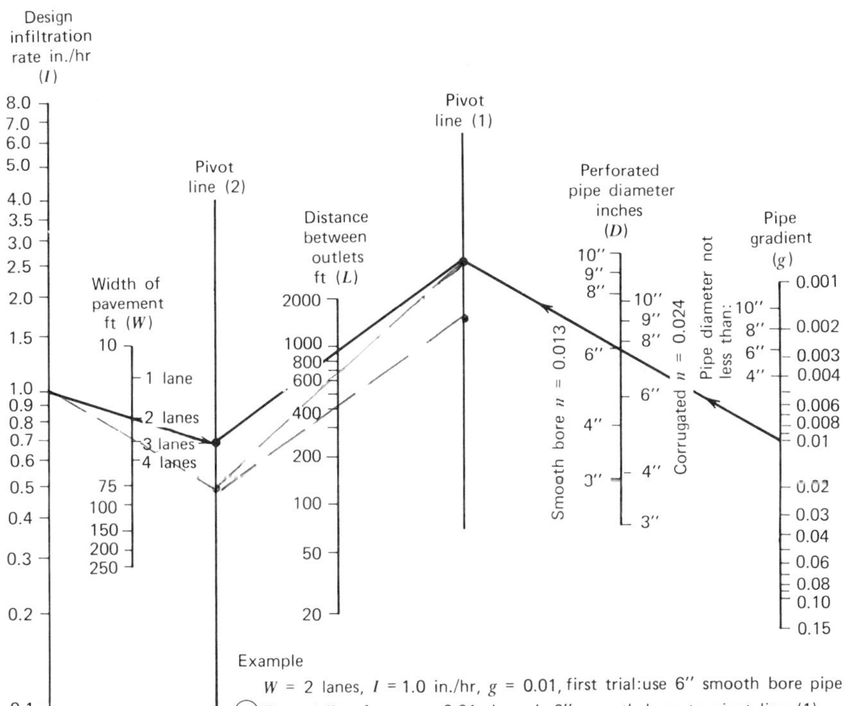


Example

$W$ = 2 lanes, $I$ = 1.0 in./hr, $g$ = 0.01, first trial: use 6″ smooth bore pipe

(1) Extend line from $g$ = 0.01 through 6″ smooth bore to pivot line (1)

(2) Extend line from $I$ = 1.0 in./hr through $W$ = 2 lanes to pivot line (2)

(3) Connect point on pivot line 1 to point on pivot line 2

(4) Read distance between outlets: $L$ = 900 ft

(5) Repeat for various pipe diameters

**FIG. 5.12** Nomograph for selection of perforated pipe diameters and outlet spacings (FHWA, 1973).

Figure 5.12 is a nomograph that was developed for easy determination of collector pipe diameters and spacing of outlets. The example on the chart explains its use (also see several applications to examples in Chapter 6). Ordinarily, the hydraulic gradient in the pipe will follow approximately the longitudinal centerline grade of a highway or airfield pavement, although in special cases it may not. For example, in areas of very flat longitudinal grades it might be more economical in some cases to steepen the grade of the pipe to permit the use of smaller-diameter pipe. In any case, the designer should try several practical outlet spacings and determine the collector pipe diameter required for each. He should then select the combination of spacing, diameter, and slope that he feels is most suitable and economical for a given project area.

## 5.9 ROAD GEOMETRICS AND OTHER RELATED FACTORS

In designing pavement-drainage systems, the designers should try to visualize these systems in three dimensions and look for any possible special or unique situations in which water could be trapped by unusual geometrics or where water would tend to meander for long distances without reaching an outlet. Some locations that require special attention are reverse superelevated curves, long sustained grades, and sag vertical curves. A suggested criterion to be used in examining pavement plans for necessary locations of collector pipes is that the maximum seepage distance should not exceed 150 to 200 ft. Longitudinal drains, transverse interceptor drains, and supplemental collector drains should be located as required to satisfy such a criterion.

Figure 5.13 is an illustration of flow paths in superelevated curves on highways. Transverse drains and edge drains are located where needed to prevent long, meandering seepage paths. Figure 5.14 shows a sag vertical curve that requires a cross drain at one or more locations, depending on the flatness of the curve and its length.

In designing subsurface drainage systems in hilly terrain, transverse drains should be installed wherever needed to prevent seepage paths from exceeding 150 to 200 ft.

When locating outlet pipes for subsurface drainage systems, designers should examine topography and look for any off ramps, structures, or other features, either natural or man-made, that will interfere with free gravity flow from the system. If suitable exits cannot be obtained at desirable locations, it may be necessary to use a longer spacing to other suitable outlet locations or to make special provisions

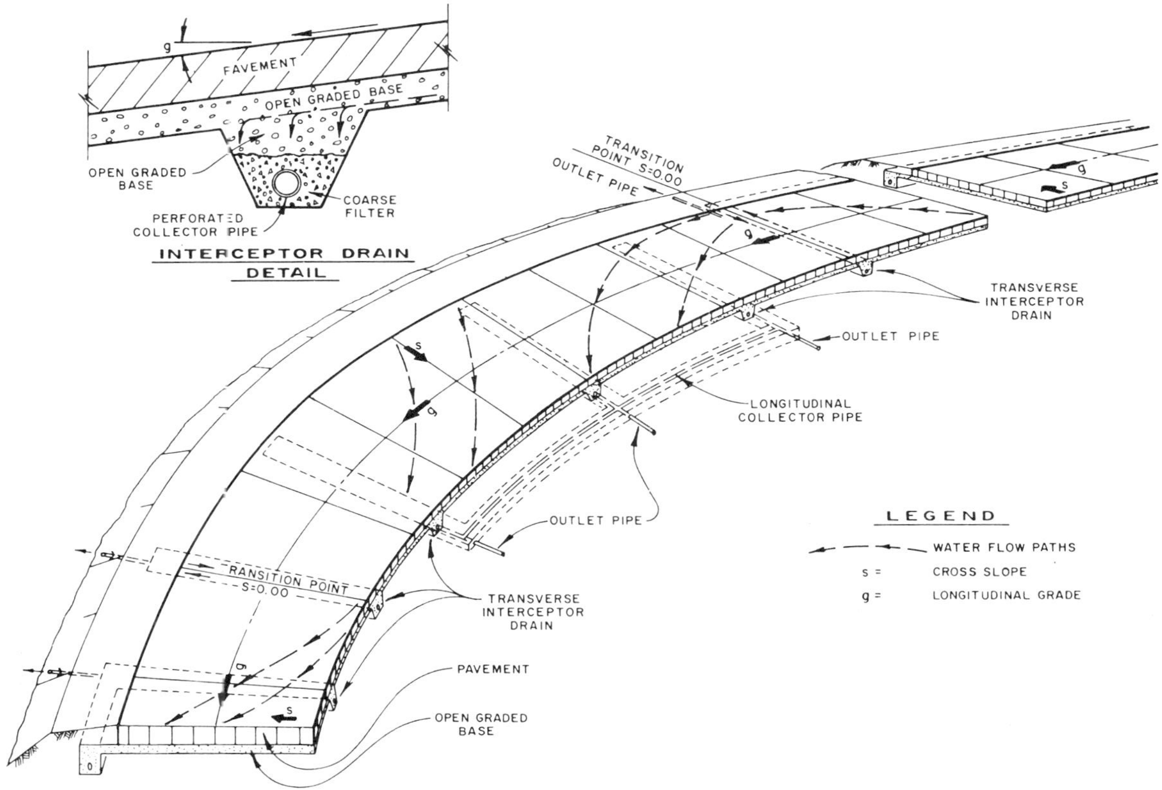


**FIG. 5.13** Transverse drains are needed on superelevated curves (FHWA, 1973).

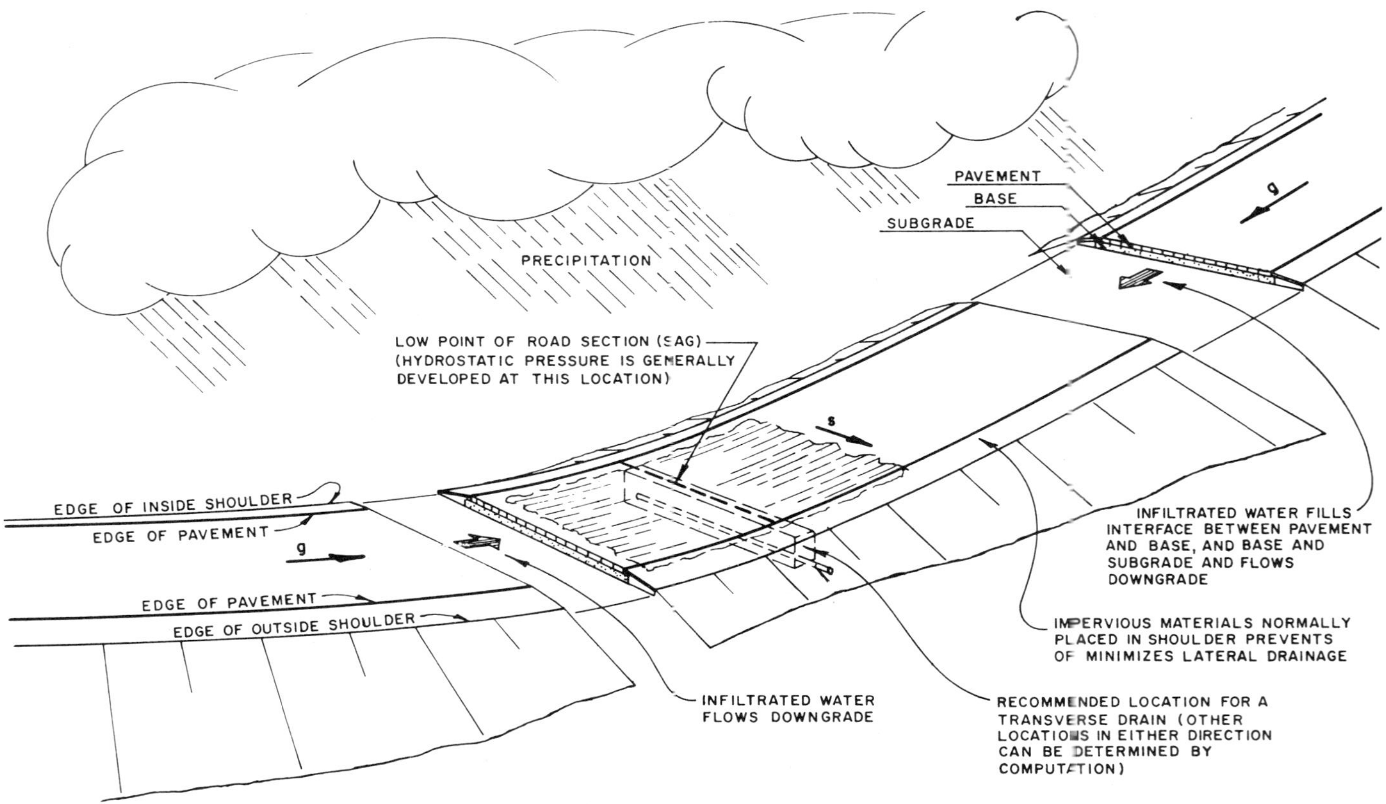


**FIG. 5.14** Transverse drains are needed on sag vertical curves (FHWA, 1973).

for sumps and pumping, although this is not likely to be required except in very unusual cases or where a road or other pavement is below the surrounding terrain and water table. In many such cases, sewers or surface-water sumps and pumps can be utilized for removal of seepage from subsurface drainage systems.

Outlets should always be located high enough on the slopes of ditches, and so forth so that free gravity flow will be assured. In unusual cases, where large flood flows cannot be avoided and the high water levels in ditches will rise on occasion above the exit pipes, flap gates should be provided (see Fig. 7.4).

The kind of material used for the pipes should be compatible with local conditions, so that it will not corrode, rust, disintegrate, or be attacked in any detrimental way by the chemical content of either the soil or the water, or any foreign matter. Standard ASTM or AASHO or federal or other agency specifications should be used.

In cold regions, collector pipes should be located below the depth of frost penetration wherever possible. When feasible, outlet pipes in cold regions should discharge into manholes, box structures, or other protective facilities and should have flap gates to keep cold air out of their ends. The gates should be designed to minimize sticking when water freezes in the inverts of the outlet pipes. Encasing outlet pipes in insulation material (Fig. 5.15) has been suggested by George W. Ring III of Washington D.C. in regions where deep freezing can be expected. Increasing the diameter of the outlet pipes above the dimensions needed for discharge quantity can provide additional discharge area to

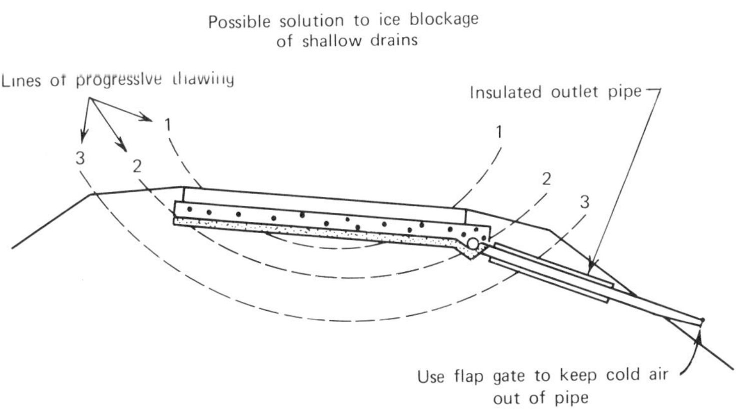


**FIG. 5.15** Insulated outlet pipe in regions having deep frost penetration (Courtesy George W. Ring).

**FIG. 5.16** Good marker posts are essential for subsurface drainage systems. (*a*) A good outlet marker post, marker, and well-located outlet pipe; a south-central state; November 1970. (*b*) A substantial outlet marker post for an interstate highway; a south-central state; November 1970.

compensate for partial reduction in flow capacity by the formation of ice in pipes.

Outlets should be clearly marked (see Fig. 5.16) so that personnel involved in safety, aesthetics, landscaping, slope erosion control, and other programs in operation of a facility are aware of the presence of the outlets and will not conduct any activity that will block or damage them or other parts of the subsurface drainage system.

## 5.10 SUMMARY COMMENTS

In designing filters and drains to remove water for the protection of engineering structures such as pavements, it is important that certain fundamental principles be applied, or the drainage systems may fail to accomplish the desired results. This chapter gives a number of important basic drainage system design criteria. Suggestions are given as to *where* subsurface drains may be needed. Also several standards for seepage performance are given, including the quantities of water that can be removed, the time needed for water particles to travel completely through the drainage layer, and the time for fall of saturation mounds after inflows stop. In addition, basic filter criteria that protect drains from clogging are given. Other design and performance standards included pertain to the protection of sensitive subgrades from surface infiltration, the permeability requirements of trench backfill material, the selection of pipe diameters and spacings for collector pipes and outlet pipes, and special considerations needed in relation to road geometrics and other factors.

When the discharge requirements and filter protection of subsurface drainage systems are given proper attention, and other basic needs are kept in mind, systems can be designed that will permanently give a high level of protection to pavements from the harmful effects of excess water.

## References

Barber, E. S. and C. L. Sawyer (1952), "Highway Subdrainage," *Public Roads,* Vol. 26, No. 12, February 1952.

Bertram, G. E. (1940), "An Experimental Investigation of Protective Filters," Publication No. 267, Graduate School of Engineering, Harvard University, January 1940.

Calhoun, C. C., Jr., J. R. Compton, and W. E. Strohm (1971), "Performance of Plastic Filter Cloths as a Replacement for Granular Filter Materials," Waterways Experiment Station, U. S. Army Corps of Engineers, Vicksburg, Miss., Presented at HRB Annual Meeting, January 1971.

Calhoun, C. C., Jr. (1972), "Development of Design Criteria and Acceptance Specifications for Plastic Filter Cloths," *Technical Report S-72-7,* U.S. Army Engineer Waterways Experiment Station, June, 1972.

Edgar, W. H. (1944), "Soil Movements as Affecting Pavement Surfaces," *Surveyor,* Vol. 103, No. 2746, September 8, 1944, pp. 427–428.

Federal Highway Administration, FHWA (1973), *Guidelines for the Design of Subsurface Drainage Systems for Highway Structural Sections.*

Morrison, R. L. (1931), "Causes of Failure of Pavements," *Highway Supervision and Construction,* Vol. 38, No. 13, April 1, 1931, pp. 34–37

Terzaghi, Karl (1922), "Der Grundbruch an Stauwerken und Seine Verhutung," *Die Wasserkraft,* Vol. 17, No. 24, pp. 445–449.

Terzaghi, Charles (1929), "Effect of Minor Geologic Details on the Safety of Dams," American Institute of Mining and Metallurgical Engineers, *Technical Publication* 215, pp. 31–44.

U.S. Army Corps of Engineers (1941), "Investigation of Filter Requirements for Underdrains," Technical Memorandum No. 183-1, Waterways Experiment Station, Vicksburg, Mississippi., December 1941.

U.S. Army Corps of Engineers (1945–1946), "Subsurface Drainage Investigation, Comprehensive Report," Office of the Chief of Engineers, Washington, D.C., authored by Engineer Office, Boston, Massachusetts

U.S. Army Corps of Engineers (1947), "Roads, Runways, and Miscellaneous Pavements," Technical Manual TM5-624, May, 1947.

U.S. Army Corps of Engineers (1965), "Drainage and Erosion Control—Subsurface Drainage Facilities for Airfields," Technical Manual TM5-820-2, August, 1965.

U.S. Bureau of Reclamation (Karpoff, 1955), "The Use of Laboratory Tests to Develop Design Criteria for Protective Filters," *Proceedings,* American Society for Testing Materials (ASTM), Vol. 55, p. 1183.

U.S. Bureau of Reclamation (1965), *Design of Small Dams,* U.S. Government Printing Office, Washington, D.C., p. 174.

CHAPTER SIX

# Typical Subsurface Drainage Systems

## 6.1 BASIC DESIGN FEATURES

Although details and general proportions, depths of ditches for collector pipes, and so forth must be adapted to the geometry and other physical conditions at individual sites as well as to the design requirements of individual pavements, some illustrative examples of drainage systems for various kinds of pavements and site conditions are given in drawings in this section. In these drawings, thicknesses of layers and sizes of drains and other components are generally exaggerated for clarity of presentation (see Fig. 6.1).

Figure 6.1*a* illustrates drainage and stabilization features for the two separate halves of a six-lane divided freeway in an area in which both rainfall and sidehill seepage need to be removed. Because of low strengths of foundation strata and heavy seepage in porous layers below the original ground surface, the designers have decided to use a deep "stabilization trench" under the fill. This trench is to have a very permeable filter blanket placed on its back slope and bottom, as shown, with pipe outlets to take the water to safe exits away from the embankment. The inside cut slope will be stabilized with an aggregate buttress fill. Roadbed drainage is provided by a fairly deep trench drain for the inside half of the freeway because it is subjected to sidehill seepage, but only a shallow V trench is needed for the outer half, because the deep stabilization trench with its drain should keep this side of the roadbed free of groundwater. Open-graded base drainage layers on suitable subbases are provided for both halves of the freeway.

If soil strata or geological formations in hilly and mountainous terrain contain highly permeable zones, layers, or joints that can become filled with water, this can in many cases add to the amount of water entering the structural sections of highways constructed in such areas. Large hydrostatic pressures that often build up in the formations in such slopes can cause serious slipouts.

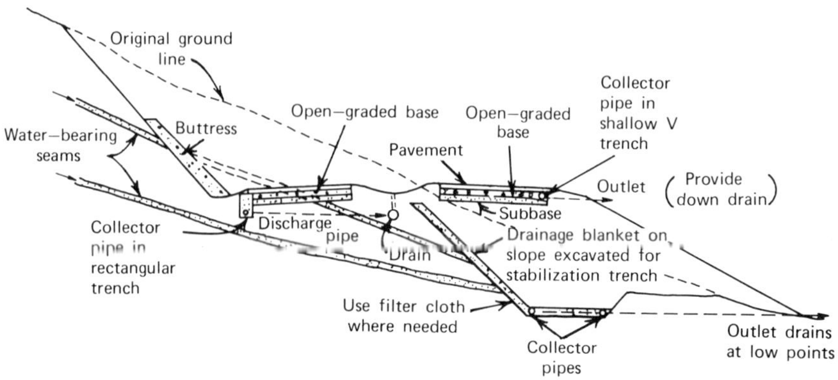


(a)

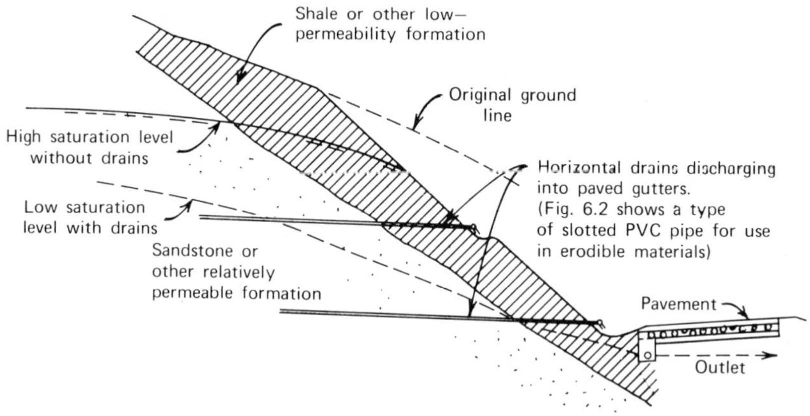


(b)

**FIG. 6.1** Drainage and stabilization features for freeways in wet climates and wet, unstable hillsides. Outside lanes have stabilization trench and shallow V drain for roadbed; inside lanes have buttress fill and trench drain; all lanes have open-graded base drainage layer. (*a*) Cross section. (*b*) Example of slope stabilization with horizontal drains.

*Horizontal drains* have in many cases been very effective for stabilizing earth slopes where conditions were such that the drains could intercept and remove the trapped water. To install this type of drain, small diameter holes are drilled horizontally into a slope needing to be drained (Cedergren, 1967), and a 1½-in. or larger diameter perforated metal or slotted PVC pipe is inserted during or after withdrawal of the drill casing. Figure 6.1*b* shows a type of geologic condition which could

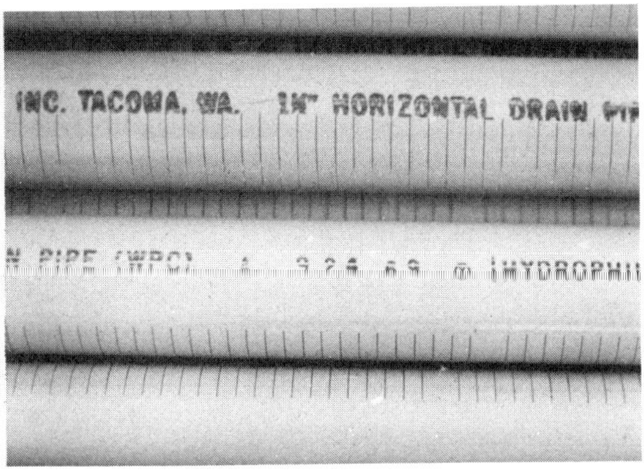


**FIG. 6.2** Slotted PVC pipe (1.4 in. I.D.) used in horizontal drain installations.

be greatly improved in stability by the use of horizontal drains. Here an outer zone of relatively impermeable material traps water in the slope of the highway, causing a high saturation level and high water pressures within the slope. If the trapped water can be effectively removed by horizontal drains, as assumed in this case, the stability of the slope can be greatly improved. Also, it may be possible to greatly reduce flows into the structural section of the road.

Figure 6.2 shows a type of slotted PVC pipe that was developed for use in horizontal drains in erodible soil formations. Slots can be cut in widths of 0.01 in. and larger to allow free flow of water into the drains while providing filter protection for nearly every kind of erodible, water-bearing soil. Soil Sampling Service Inc. of Puyallup, Washington holds U.S. Patent No. 3,391,543 covering processes for installing drains with this kind of pipe.

Figure 6.3 is a cross section through an airfield runway or taxiway in flat terrain (as in other examples, vertical dimensions and sizes of

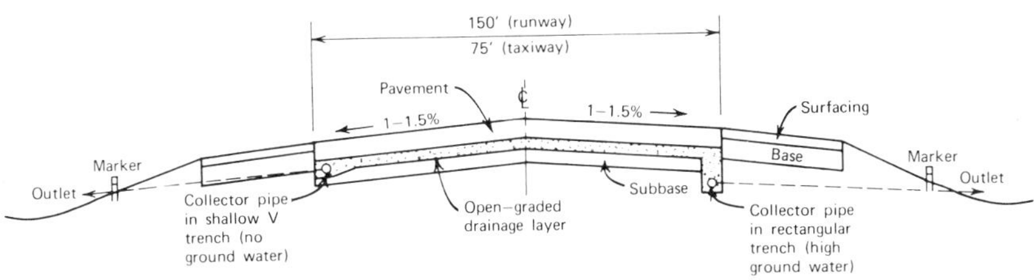


**FIG. 6.3** Drainage system for airfield runways and taxiways.

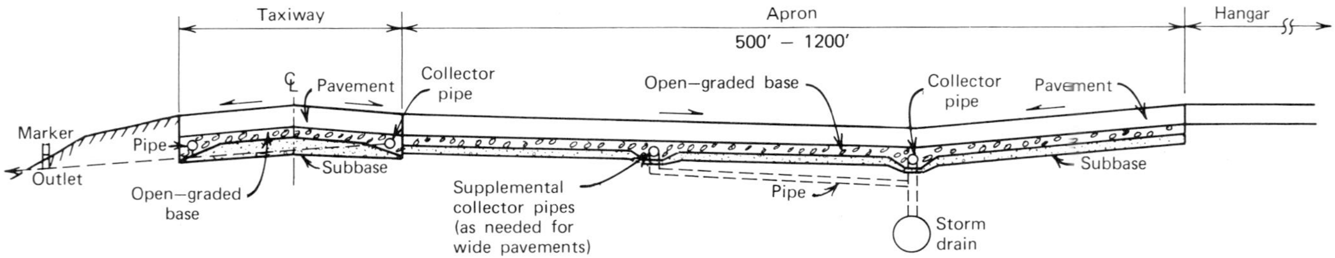


**FIG. 6.4** Drainage system for airport parking apron and adjacent taxiway.

drains have been exaggerated for clarity). A drainage blanket is provided under the full width of the runway or taxiway, and pipe drains are located under the two outer edges, as shown, with outlets discharging into shallow swales at the sides of the pavements. Low marker posts call attention to the outlet locations.

A wide airport parking apron is depicted in Fig. 6.4, adjacent to a heavy-duty taxiway. The collector drains for the taxiway discharge to a natural swale to the left; however, interior collector pipes for the wide apron discharge into a storm sewer pipe located under the apron. The subsurface drainage systems for very wide airport pavements such as these must be adapted to the conditions at individual sites, utilizing natural drainage swales, constructed ditches, storm drains, or special sewer pipes when needed to achieve positive drainage from the systems.

Figure 6.5 shows a depressed freeway section with an open-graded base drainage layer discharging to longitudinal collector pipes that empty into a sump from which the water is removed by pumps that also remove surface water that enters the depressed section. A float actuates the pumps as needed to keep seepage or runoff from raising the water in the sump above a safe level.

Figure 6.6 is a cross section of a city street or highway with a subsurface drain that discharges into underlying storm sewers. This is often the most practical way to remove water from drainage systems where the adjacent land is highly developed and drainage by gravity is not possible. In many such cases storm drains are readily available for removal of the subsurface water from streets, parking lots, and so forth.

Figure 6.7 illustrates details at the outer edges of subsurface drains for pavements that have shoulders along their outer, lower edges. The manner in which the design should be handled in such areas depends

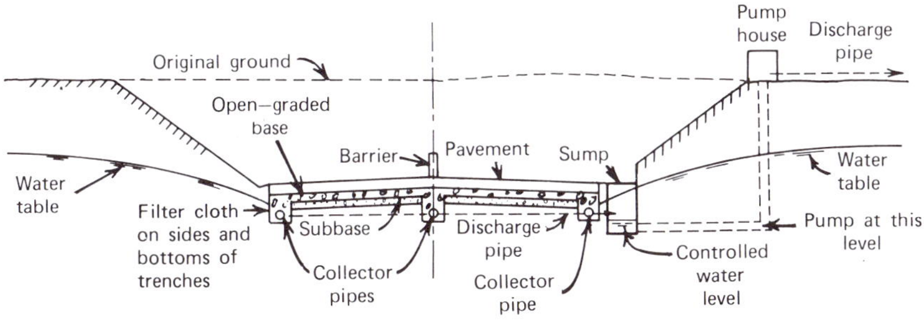


**FIG. 6.5** Drainage system for freeway below the normal water table. Seepage discharges into storm drainage system.

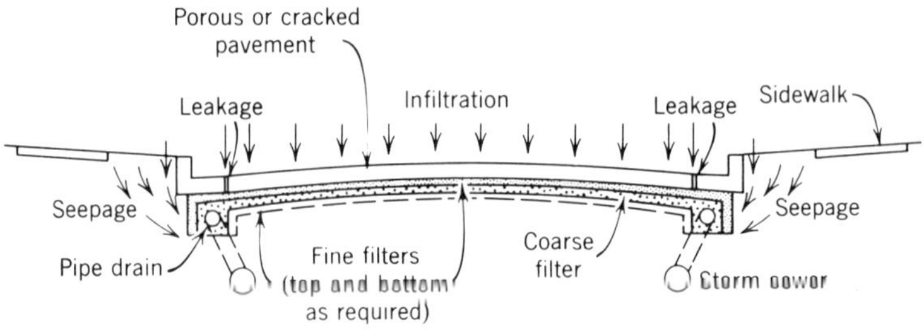


**FIG. 6.6** Drainage system for city streets or highways subjected to heavy infiltration and seepage. Discharges into storm sewers ("Seepage, Drainage, and Flow Nets," John Wiley and Sons, New York, 1967, p. 392).

in part on the usage of the pavements and shoulders. Freeways that are constructed of PCC pavements with AC shoulders often have serious deterioration taking place along the joint between the PCC and the AC (see Fig. 3.1*c* and *d*). The breaking down of this joint is often caused by heavy trucks riding out on the shoulder (particularly around horizontal curves), together with thermal expansion and contraction of the primary pavement, which tends to open this joint, and other actions such as freezing and thawing, the entry of water, pumping of the adjacent pavement under heavy wheel loads, and so forth. To protect this joint area, the longitudinal drain pipe can be located slightly outside of the joint, as shown in Fig. 6.7*b*, or if it is advisable to protect the entire shoulder, the drainage layer can be extended, and the longitudinal collector drain located at the outer edge, as shown in Fig. 6.7*c*.

For airfield pavements that will have no heavy wheel loads at or near this joint or on the shoulder, the outer longitudinal drain can often be located under the outer edge of the primary pavement, as shown in Fig. 6.7*a*.

If the pavements are in cold regions where frost damage to shoulders can occur, the drain can be located under the outer edge of the shoulder, with a shallow "V" trench being used as in Fig. 6.7*c* if frost penetration is shallow, or a deeper rectangular trench as shown in Fig. 6.7*e* if frost penetration is too deep for the shallower V trench.

In many situations in which there is no groundwater and no appreciable frost penetration, shallow V ditches will suffice for the longitudinal collector drains, but where there is high groundwater or deep frost penetration, trenches should be excavated to sufficient depth to provide the necessary lowering of the saturation level to prevent damage to the pavements (Fig. 6.7*d* and *e*).

Where trench drains are needed, their design should be developed to provide the required seepage control at the least cost. Figure 6.8*a* shows a trench with dimensions normally used, whereas the one in Fig. 6.8*b*, which was suggested by Gordon Benson and Donald Fowler of the Illinois Department of Transportation, uses an inexpensive type of agricultural pipe in a very narrow trench. Both designs use pea gravel, stone chips, or other very permeable backfill material, so as not to inhibit the free flow of water to the pipes. Unfortunately trench drains are often

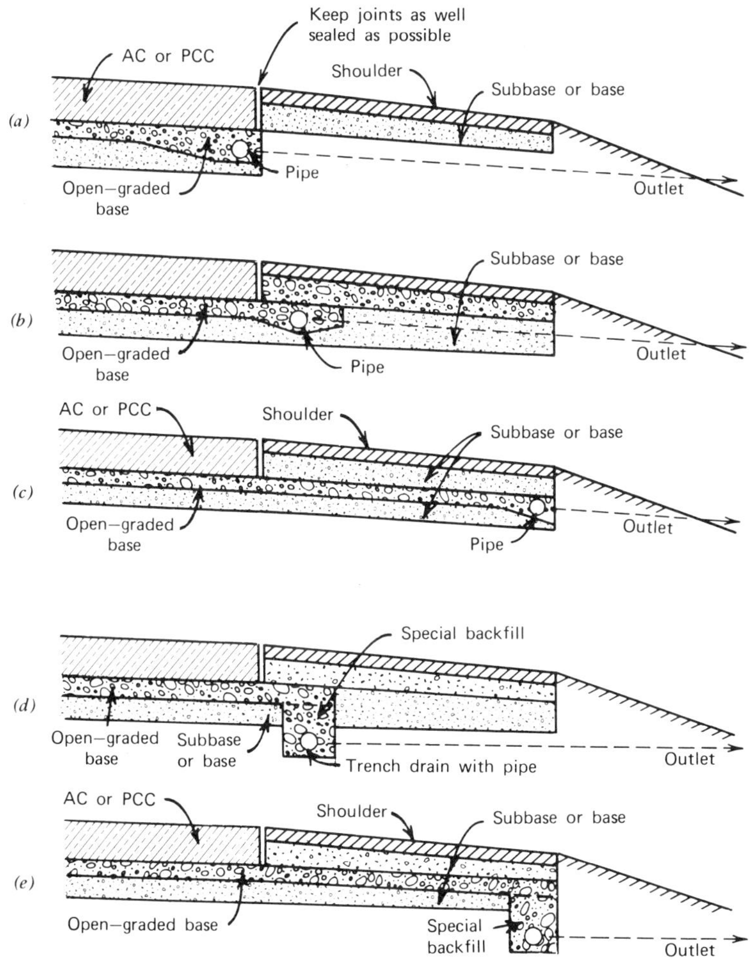


**FIG. 6.7** Typical details of outer edges of drainage systems.

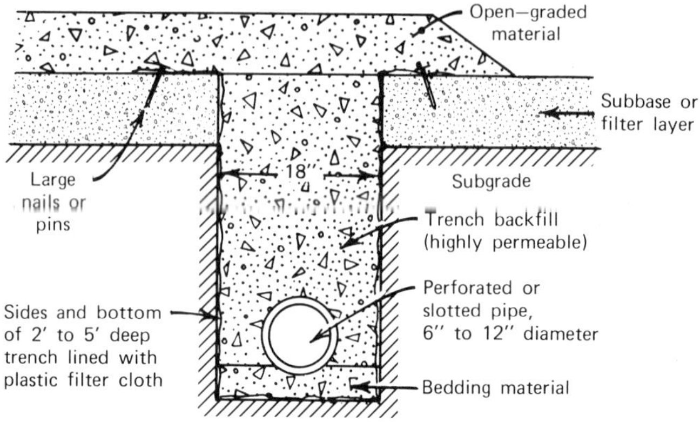


(*a*) Normal dimensions

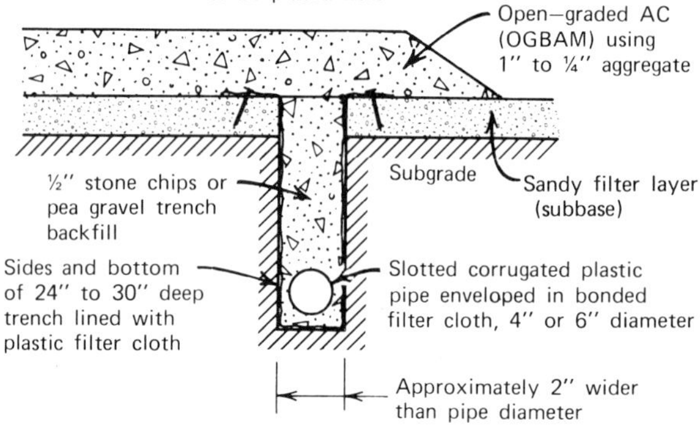


(*b*) "Mini—trench" design. Suggested by Gordon Benson and Donald Fowler, Illinois Dept. of Transportation, 1973.

**FIG. 6.8** Details of trench drains with plastic filter cloths on sides and bottom. Two designs.

backfilled with low-permeability sand and gravel blends that severely restrict transmissibility and prevent water from getting to the pipes.

As already noted, whenever an open-graded base drainage layer is used for roadbed or airfield pavement drainage, a base or subbase that is placed between this layer and a soil subgrade should have a gradation that will provide filter protection to prevent the intrusion of fine, erodible soil materials into the open-graded layer. If a trench drain is cut into erodible soils, plastic filter cloth can be placed on the sides and bottom of the trench to keep soil out of the backfill material as shown in Fig. 6.8.

Figure 6.9 shows gradations and permeabilities of several open-graded aggregate drainage materials and several blended aggregates in the class of filter materials. They are all high quality, washed, processed aggregates containing moderate percentages of crushed particles. All of the open-graded materials were stabilized by hot-mixing with a small amount of a paving grade asphalt. Three are similar to those for which tests are summarized in Table 6.1. The filter materials have been compacted to only moderate densities.

The permeability values in Fig. 6.9 illustrate general levels of permeability that are possible for the range of material gradations depicted. While open-graded materials of the kind represented are only slightly affected by increased compactive effort, blends of sand and gravel sizes are greatly influenced by compactive effort and water

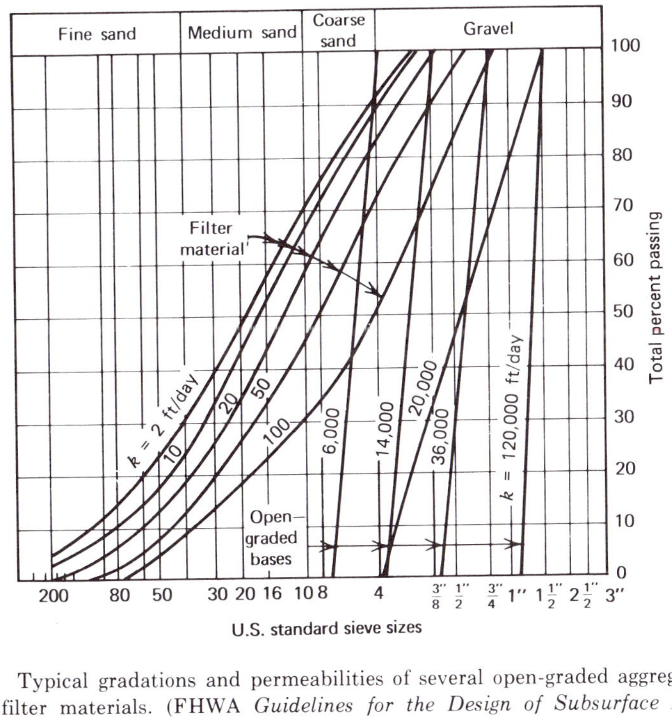


**FIG. 6.9** Typical gradations and permeabilities of several open-graded aggregates and several filter materials. (FHWA *Guidelines for the Design of Subsurface Drainage Systems for Highway Structural Sections,* Figure No. 4).

content during compaction. Consequently, the filter materials shown in Fig. 6.9 could have much lower permeabilities than the values given, depending on the state of compaction obtained in the work.

Materials with the levels of permeability needed in the open-graded drainage layers for pavements are distinctly different from most of the finer types of road building materials in widespread use. Leaving out the fines gives the open-graded materials a visibly open, porous appearance, in contrast with the nonporous appearance of most fine materials. The smooth, tight texture of some commonly used materials has been shown in photographs in other chapters (e.g., the sand subbase shown in Fig. 4.21*b*). The photographs in Fig. 6.10 show the striking differences between fine and coarse materials. Figure 6.10*a* is of a sandy subbase and a gravelly sand base course under a PCC wearing course used in 1970 in the construction of an interstate highway in a state bordering one of the great lakes. The excavation at the end of a completed portion of this pavement shows the compact, dense appearance of both of these layers. Fine-grained blends of sand and gravel such as these often have coefficients of permeability in the range of 0.1 to 10 ft/day ($3 \times 10^{-5}$ to 0.003 cm/sec).

Two types of material of much higher permeability are shown in Fig. 6.10*b*. The one on the right is a very clean sandy gravel that grades from $\frac{3}{4}$ in. maximum size to a few percent passing a no. 20 sieve and has a coefficient of permeability of about 100 ft/day (this could be a good filter material, but is not permeable enough for an open-graded drainage layer). The sample on the left is a hard, broken stone grading from $1\frac{1}{4}$ to $\frac{1}{2}$ in. size with a coefficient of permeability in excess of 100, 000 ft/day (35 cm/sec or higher). Figure 6.10*c* shows the $\frac{3}{4}$ to $\frac{3}{8}$ in. material (asphalt treated) included in Table 6.1 (coefficient of permeability = 35,000 ft/day). This aggregate, although only moderately coarse, has good water transmitting capabilities, and a 2-in.-thick layer has more than 500 times the drainage capability of a 12-in.-thick layer of sandy gravel with a coefficient of permeability of 10 ft/day!

Figure 6.10*d* shows pieces of hard, durable crushed rock grading from 2 to $\frac{1}{2}$ in. in size that were obtained by diamond coring into a base under a 6-in.-thick AC surface course at a major airfield in an investigation sponsored by the Construction Engineering Research Laboratory (CERL) of the U.S. Army Corps of Engineers (1974). The strong interlocking allowed large pieces to be completely cored through without becoming dislodged. This type of aggregate is not only an excellent drainage material but one of the best, most economical materials from a structural design standpoint.

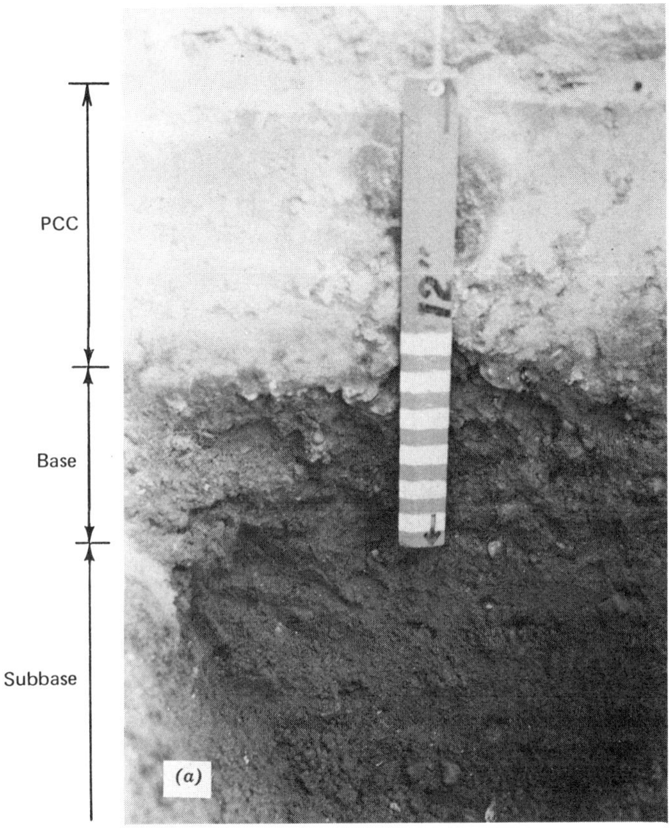


**FIG. 6.10** Photos showing general appearance of some low and high permeability materials. (*a*) Exposed 4-in. gravelly sand base and sandy subbase under PCC pavement on an interstate highway under construction. All low permeability materials. Great Lakes area. October 1970. (*b*) Two samples of permeable aggregates; on right, very clean sandy gravel, $k$ = 100 ft/day; on left, coarse broken rock, $k$ = 100,000 ft/day or more. (*c*) Asphalt-stabilized gravel, $\frac{3}{4}$- to $\frac{3}{8}$-in. material, $k$ = 35,00 ft/day. (*d*) Hard, durable trap rock shows coring marks of diamond drill used in taking sample of airfield taxiway pavement; an excellent type of base; pavement in excellent condition.

**FIG. 6.10** (Continued).

## 6.2 DRAINAGE MATERIALS

Criteria for checking the gradation and permeability requirements of aggregate drainage layers are given in Chapter 5. All materials used in drainage systems should be tested by standard methods to determine their gradations, durability, and (when necessary) the permeability. All samples used for determination of compliance with specification limits or requirements should be taken after the materials have been placed and compacted in the work, so that the tests will be representative of the actual *in-place* materials. Tests should be made using procedures and apparatus that will ensure reliable results. Errors in permeabilities can occur if restrictions in flow are induced through the use of fine filter papers or tubes or hoses of small diameters, by air-locking when tap water is used, and other improper procedures.

Flow in extremely coarse drainage layers is likely to be semi-turbulent in some cases; hence Darcy's law may sometimes tend to overrate the discharge capacities of these layers. If the hydraulic gradients applied in the tests are about equal to those that will develop in the drainage layers, the effects of turbulence may be largely compensated for in the tests.

In Chapters 4 and 5 it is shown that if drains are to provide high levels of protection to pavements from excess water, the drainage layer (open-graded macadam layer) must have a coefficient of permeability of thousands of feet per day. To achieve high levels of permeability, the drainage material should be one-sized or of a narrow range of sizes and must not contain any significant quantity of fine particles that can clog pores and restrict permeability (see Fig. 5.1, and Eqs. 5.1 and 5.2). To minimize segregation during placement and other possible construction difficulties, these layers generally should not contain any particles larger than about 1 in. or possibly 1½ in. maximum size. Within the recommended range of aggregate sizes, coefficients of permeability will generally be in the range of 5000 to over 100,000 ft/day (1.7 to 35 cm/sec), permitting designs to be developed for a wide range of transmissibilities. Some permeabilities that were obtained in laboratory tests on materials of this kind are given in Table 6.1, both without any additives and as stabilized by hot mixing with about 2 percent of a paving-grade asphalt.

Including even small amounts of finer sizes in drainage aggregates can greatly restrict permeability. Childers (1968) made a study of gradations of drainage materials in relation to permeability and stability. Both permeability and CBR tests were made on samples having varying amounts of material passing the no. 10 sieve. This study indicated that to maintain a permeability of at least 100 ft/day, compacted crusher run materials should not contain more than 20 percent of particles passing the no. 10 sieve. When blends of sand and gravel containing only small amounts of silt or clay are compacted to the

**TABLE 6.1** Laboratory Permeabilities of Untreated and Asphalt-Stabilized Open-Graded Aggregates

| Aggregate size range | Permeability (ft/day) | |
|---|---|---|
| | Untreated | Bound with 2% asphalt |
| $1\frac{1}{2}$ to 1 in. | 140,000 | 120,000 |
| $\frac{3}{4}$ to $\frac{3}{8}$ in. | 38,000 | 35,000 |
| No. 4 to no. 8 | 8,000 | 6,000 |

(Lovering and Cedergren, 1962)

degree required in structural layers, their permeabilities can be drastically reduced. Strohm, et al. (1967) found that when such materials are thoroughly compacted, the effective porosities can approach zero with as little as 5 percent of fines (material passing a no. 200 sieve), and the permeabilities are relatively infinitesimal.

The *Guidelines* (FHWA, 1973a) recommends that " . . . a layer of highly permeable base (be) placed immediately beneath paved areas requiring protection from excess water." It also says, "In no case should open-graded base sizes be smaller than No. 4 sieve, since the smaller size materials greatly reduce permeability." The *Final Report* in the study for the development of these guidelines (FHWA, 1973b), says that,

> Crude estimates or even educated "guesses" of inflow rates, if tempered with experience and judgment, will almost always indicate the need for drainage systems that contain a layer of one-sized stone of pea gravel or larger size. Even a very minimal full-width dual-layer subdrainage system containing a few inches of highly permeable stone can remove hundreds, if not thousands of times as much water as thicker layers of most of the so-called "pervious" bases in use today.

The provision of such a system will not only ensure a high level of protection from surface water infiltration, but will give a margin of safety against random or unknown inflows from springs and so forth.

The term "dual-layer" means an open-graded layer with a protective filter. An indication of the high levels of permeability needed in these pavement drainage systems is given by Fig. 6.11, which compares the capabilities of several open-graded materials to remove surface water infiltration with some commonly used base materials.

As a guide in making tentative selections of open-graded base drainage materials, several typical designs are presented in Table 6.2, with comments about the general kinds of conditions for which each might be considered. Actual designs for projects should be verified by making calculations as shown in several examples in Sec. 6.3. If probable inflows can be only roughly estimated, designers can provide large margins of safety by requiring relatively thick layers of very permeable materials.

When designers have a choice of several gradations of open-graded aggregates for drains at about the same unit prices, greater economy is usually obtained by selecting relatively coarse materials. Using coarser, more permeable material permits a thinner drainage layer to be used to obtain a given amount of drainability or transmissibility—subject to

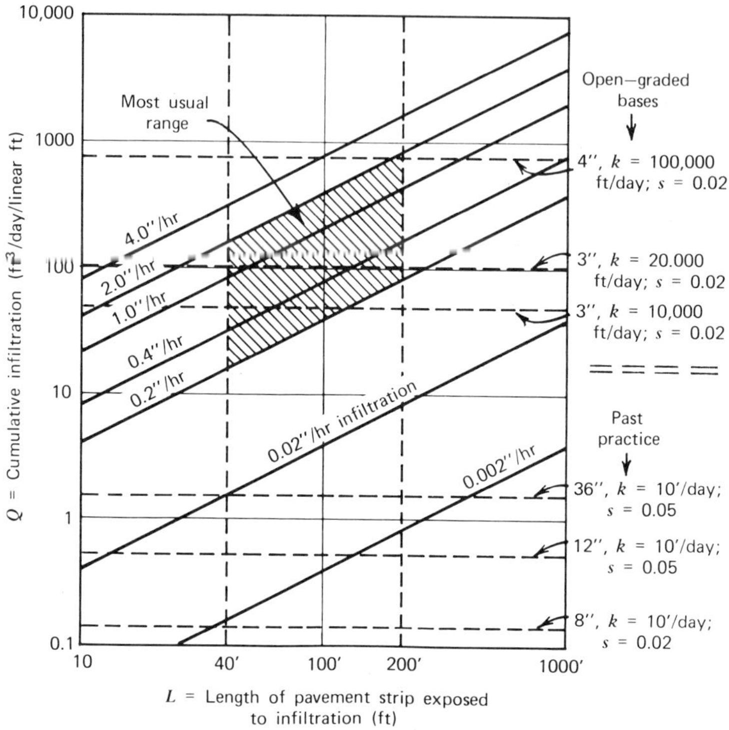


**FIG. 6.11** Chart illustrating need for high permeabilities in base drainage layers.

construction feasibility limiting the minimum thickness that it is practical to place. Extremely thin layers of aggregate are often spread in the construction of armor coats, chip seals, and so forth; however many designers feel that drainage courses should be at least a few inches thick.

Sometimes, relatively thick drainage layers of about the finest particle sizes within the suggested range may be used to obtain the required transmissibility, particularly when there is a structural advantage in using the thicker drainage layer or if coarser materials are scarce or costly in a given area. The most "optimum" design for each case will depend on numerous factors.

## 6.3 TYPICAL DRAINAGE ANALYSIS CALCULATIONS

Several examples are given here to illustrate procedures that can be used in designing subsurface drainage systems that will be adequate in

**TABLE 6.2** General Capabilities of Several Open-Graded Drainage Layers (A Preliminary Guide)

| | Open-graded drainage layer | | |
|---|---|---|---|
| Example | Thickness (in.) | Aggregate size (in. or no.) | General suitability |
| *A* | 3 to 4 | $\frac{3}{8}$ to $\frac{1}{8}$ | A minimal design of relatively fine material; areas of low rainfall and small inflows from other sources; keep joints and cracks and porous pavements well sealed; for narrow pavements |
| *B* | $2\frac{1}{2}$ to 3 | $\frac{3}{8}$ to No. 4 | A minimal thickness design for regions where coarse aggregates are scarce or costly; no significant groundwater, springs, etc.; keep pavements well sealed; for narrow pavements |
| *C* | 6 to 9 | $\frac{3}{8}$ to $\frac{1}{8}$ | A moderate design of relatively fine material; areas of moderate rainfall and minor inflows from other sources; keep pavements well sealed; moderate widths of pavements |
| *D* | $2\frac{1}{2}$ to 3 | 1 in. to $\frac{1}{2}$ | A minimal thickness design of moderately high capacity; for areas of moderate rainfall and moderate inflows from other sources; keep pavements well sealed; moderate widths |
| *E* | 3 to 4 | $\frac{3}{4}$ to $\frac{3}{8}$ | A moderate design; use in areas of moderate rainfall and moderate additional inflows; keep pavements as well sealed as possible; moderate widths |
| *F* | 5 to 6 | 1 in. to $\frac{1}{2}$ | Heavy capacity design; for areas of heavy rainfall; moderate additional inflows; keep pavements well sealed; for rather wide pavements |
| *G* | 8 to 9 | $1\frac{1}{4}$ in. to $\frac{3}{4}$ | Heavy capacity design; heavy inflows; rather wide pavements; keep pavements well sealed |

Note: Filter criteria for adjacent layers must be satisfied, and open-graded layers must be protected from infiltration and clogging both during construction and under maintenance. Drainage systems should be designed to meet individual site conditions; this table is a preliminary guide only.

terms of: (1) *transmissibility* needed to remove estimated inflow rates and (2) *permeability* needed to allow water to flow through with sufficient rapidity to minimize freezing and other damage. Solutions are developed both by direct calculations with Darcy's law and with the aid of several charts that facilitate these determinations.

**Example 1 Interstate Highway (Three Lanes Each Direction).** This highway, in steep terrain, has a longitudinal grade of 5 percent (0.05) and a cross slope of 2 percent (0.02); its design precipitation rate is 1.5 in./hour, and the design infiltration rate is ⅔ the design precipitation rate or (0.67) (1.5) = 1.0 in./hour; no other inflows; no significant shoulder or subgrade drainage; erodible subgrade (see Fig. 6.12). Effective porosity of drain = 0.2.

First, use Darcy's law to analyze flow conditions in a 1-ft-wide strip of pavement and open-graded base drainage layer in the direction of flow. As may be seen in Fig. 6.12, the length of such a flow path is 108 ft, and the gradient in the direction of flow is 0.054. Then, from Darcy's law, the required *transmissibility* $kA$ to drain the infiltration rate of 1.0 in./hour = $kA = Q/i$. In this example, the infiltration per square foot = $q$ = 1.0 in./hr (24/12) = 2.0 ft/day, and the total quantity of water entering the strip is $Q = 108q$; so $Q = 108\ (2.0) = 216$ ft³/day. Then, $Q/i = 216/0.054 = 4000$ ft³/day; hence $kA = 4000$ ft³/day. The drainage layer must have at least this transmissibility.

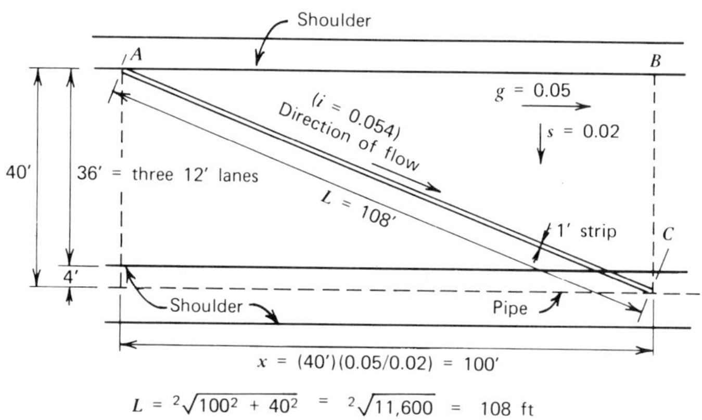


$L = \sqrt[2]{100^2 + 40^2} = \sqrt[2]{11{,}600} = 108$ ft

$\Delta h_{A-B} = 100\ (0.05) = 5.0$ ft

$\Delta h_{B-C} = 40\ (0.02) = 0.8$ ft

$\Delta h_{A-C} = 5.8$ ft

$i = 5.8/108 = 0.054$

**FIG. 6.12** Data for Darcy's law calculations for drain design. Example 1—interstate highway.

TRIAL 1. Assume a base drainage layer thickness of 5 in. with an effective thickness of 4 in.; then $A = 0.33$ ft(1.0 ft) $= 0.33$ ft$^2$ and $k = kA/A = 4000/0.33 = 12{,}000$ ft/day.

TRIAL 2. Assume a base drainage layer thickness of 3 in. with an effective thickness of 2 in., for $A = 0.167$ ft(1.0 ft) $= 0.167$ ft$^2$. Then $k = kA/A = 4000/0.167 = 24{,}000$ ft/day.

Assume that from a structural design standpoint a 5-in. base drainage layer is more desirable than a lesser thickness, so a thickness of 5 in. will be used (effective thickness = 4 in.).

CHECK THE REQUIRED $k$ USING THE CHART IN FIG. 5.4.

For this example, $W/s = 40/0.02 = 2000$, so enter Fig. 5.4 for $W/s = 2000$, and $I = 1.0$ in./hour, and read the coefficient of transmissibility $C = k_b t_b = 50{,}000$ inch-feet/day. (Note that the ratio $W/s$ is also equal to the ratio of the seepage distance $L$ divided by the slope in the direction of seepage $= L/i = 108/0.054 = 2000$). For the assumed $t_b = 4$ in., the required value for $k_b = 50{,}000/4 = 12{,}500$ ft/day. This is a good check of the value obtained with Darcy's law for an effective thickness of 4 in.

CHECK THE REQUIRED $k$ FOR A DRAINAGE TIME OF 1 HOUR.

Use Fig. 5.6 as follows: Find family of curves for three lanes (along top side of chart); then for a longitudinal grade of 0.05 and an effective thickness of 4 in. for the drainage layer, a coefficient of permeability of 9,500 ft/day is read at the bottom of this chart. This is less than is required for transmissibility, so it is not the controlling criterion.

RECOMMENDED DESIGN OF OPEN-GRADED DRAINAGE LAYER.

Use a 5-in.-thick layer of open-graded aggregate with a minimum coefficient of permeability of 15,000 ft/day on a suitable base or subbase. No special filter is needed if the base or subbase meets filter criteria (Sec. 5.5). Referring to Fig. 5.1, it is seen that in order to have a permeability of at least 15,000 ft/day, the open-graded material should have a 15 percent size ($D_{15}$) in the order of 0.25 to 0.3 in. or slightly larger.

CHECK SIZE OF DRAIN PIPE AND OUTLET PIPE SPACING.

Use the nomograph in Fig. 5.12, and find the following values: gradient = 0.05, trial pipe diameter = 6 in. (CMP) or 5 in. (smooth bore), and $I$ = 1.0 in./hour. Lay a straight-edge between the pipe gradient = 0.05 and the trial pipe diameter = 6 in. (CMP) to pivot line 1; connect another straight-edge between design infiltration rate = 1.0 in./hour and width of pavement = 40 ft to pivot line 2; then connect between the two pivot lines to read the distance between outlets = 600 ft. This is a relatively large spacing, which suggests there is no need to try other pipe sizes; to be conservative, the outlets should probably be spaced at intervals of not over 400 to 500 ft.

**Example 2 Airport Taxiway (Half-width = 37.5 ft).** Assume a longitudinal grade of 0.002 and a cross slope of 0.02; a design precipitation rate of 2.5 in./hour; a design infiltration rate = 0.5(2.5) = 1.25 in./hour = 1.25(24/12) = 2.5 ft/day; no other significant inflows; no significant shoulder or subgrade drainage; erodible subgrade; cold climate (see Fig. 6.13). Porosity of drain = 0.3.

First, use Darcy's law to analyze flow conditions for a 1-ft-wide strip of pavement and open-graded base drainage layer in the direction of flow. As may be seen in Fig. 6.13, the length of the flow path is 40.2 ft. In consideration of the very flat longitudinal grade, assume the hydraulic gradient is equal to the cross slope = 0.02. Then, according to Darcy's law, the required *transmissibility* to remove the infiltration rate of 2.5 ft/day = $kA = Q/i$. In this example, the unit inflow = $q$ = 2.5 ft/day, and the total $Q = qL$ = 2.5(40.2) = 100 ft$^3$/day, so $Q/i$ = 100/0.02 = 5000 ft$^3$/day = $kA$, the transmissibility required in the drain.

TRIAL 1. Assume the open-graded base drainage layer has a thickness of 6 in. with an effective thickness of 5 in., for $A$ = 5/12 = 0.42 ft$^2$. Then, $k = kA/A$ = 5000/0.42 = 12,000 ft/day.

TRIAL 2. Assume the drainage layer has a thickness of 4 in. with an effective thickness of 3 in., for $A$ = 3/12 = 0.25 ft$^2$. Then $k = kA/A$ = 5000/0.25 = 20,000 ft/day.

Assume that from a structural standpoint there would be no advantage in using a base drainage layer less than 4 in. thick; then:

CHECK THE REQUIRED $k$ USING FIG. 5.4.

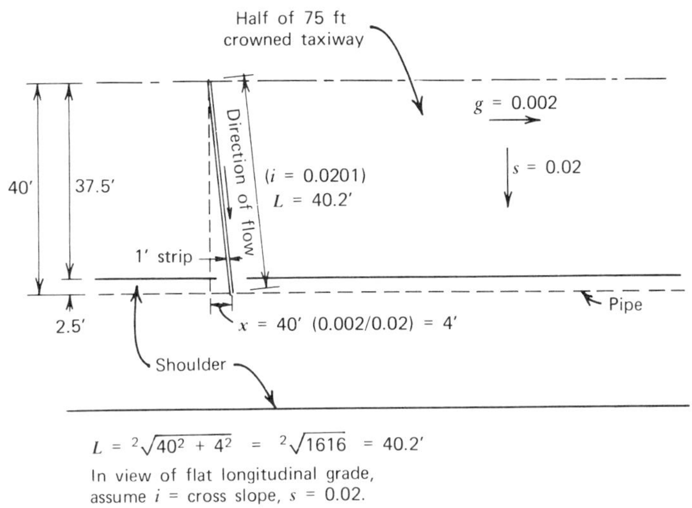


**FIG. 6.13** Data for Darcy's law calculations for drain design. Example 2—airport taxiway.

For this example $W/s = 40.0/0.02 = 2000$, so enter Fig. 5.4 for $W/s = 2000$ and $I = 1.25$ in./hour, and read the coefficient of transmissibility $C = k_b t_b = 65{,}000$ inch-feet/day. Then, since $t_b = 3$ in., the required $k_b = 65{,}000/3 = 22{,}000$ ft/day. This is an adequate check of the value obtained by direct calculation with Darcy's law for an effective thickness of 3 in.

CHECK THE REQUIRED $k$ FOR A DRAINAGE TIME OF 1 HOUR.

Use Fig. 5.6 to make an approximate check as follows: The half-width of 37.5 ft is nearly the same as three highway lanes of 12 ft, so use the family of curves for 3 highway lanes, longitudinal grade = 0.002, cross slope = 0.02, and a 3-in. effective thickness of the open graded drainage layer. For these values, the required permeability $k$ = 8500 ft/day is estimated from this chart. This is less than required for transmissibility, so it is not the controlling criterion. Another check of the required permeability can be made with Fig. 5.7. Since the curves in Fig. 5.7 are for zero longitudinal slope, and the taxiway has a grade of 0.002, very little error will be caused by this difference. By using the line for $s = 0.02$ and $t = 0.5$ ft, a permeability of approximately 5000 ft/day is estimated from this chart. Since Fig. 5.7 is for a drainage time of 2 hours, permeability values obtained from this chart must be multiplied by 2.0 to obtain values for a 1-hour drainage time. For this example, the permeability would need to be approximately 5000(2.0) = 10,000 ft/day, which is a little higher than was obtained using Fig. 5.6. This also is less than required for transmissibility.

RECOMMENDED DESIGN OF OPEN-GRADED DRAINAGE LAYER.

Use 4-in.-thick open-graded base with a minimum $k$ of approximately 25,000 ft/day. Place the open-graded base on a suitable base or subbase that will meet filter criteria.

CHECK SIZE OF DRAIN PIPE AND OUTLET PIPE SPACING.

Use the nomograph in Fig. 5.12 and find the following values: longitudinal gradient of pipe = 0.005 (trench is dug steeper than pavement grade), and a trial pipe diameter = 8 in. (CMP) or 6 in. (smooth bore), and design infiltration rate = 1.25 in./hour. Lay a straight-edge between the pipe gradient = 0.005 and the trial pipe diameter = 8 in. (CMP), to pivot line 1; connect another straight-edge between design infiltration rate = 1.25 in./hour and width of pavement = 37.5 ft to pivot line 2; then read the distance between outlets = 400 ft.

The values for pipe slope, diameter, and outlet spacing all appear reasonable; however in a detailed design analysis the relative costs of these and other choices of slopes and dimensions should be checked.

**Example 3 Interstate Highway (Three Lanes in Each Direction).** This divided highway is in low-lying, flat terrain; longitudinal grade = 0.005; cross slope = 0.02; groundwater inflow from soil with $k$ = 10 ft/day and upward hydraulic gradient $i$ = 0.5; erodible soil subgrade; design precipitation rate = 1.5 in./hour and design infiltration rate = 0.67(1.5) = 1.0 in./hour; temperate climate, no freezing (see Fig. 6.14). Porosity of drain = 0.2.

First, use Darcy's law to analyze flow conditions in a 1-ft-wide strip of pavement and open-graded base in the direction of flow. As may be seen in Fig. 6.14, the length of such a flow path is 45.5 ft, and the hydraulic gradient is 0.021. With Darcy's law, the required $kA$ to drain the design infiltration rate of 1.0 in./hour plus the groundwater inflow can be estimated as follows: the $Q$ from infiltration = (1 in./hour)(24/12)(45.5 ft) = (2 ft/day)(45.5) = 91 $ft^3$/day. The

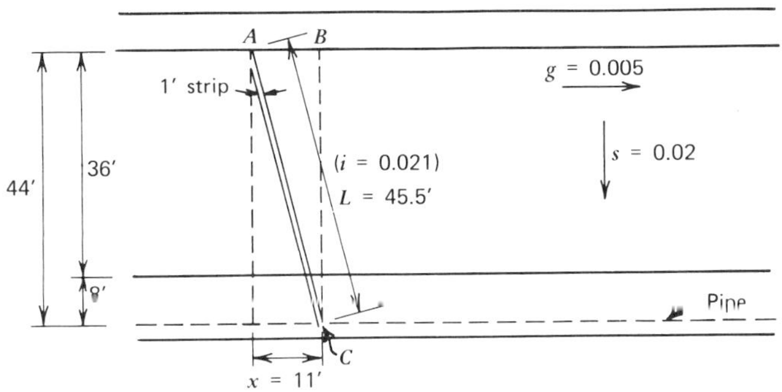


$$x = 44' \ (0.005/0.02) = 11 \text{ ft}$$

$$L = \sqrt[2]{44^2 + 11^2} = \sqrt[2]{1936 + 121} = \sqrt[2]{2057} = 45.5 \text{ ft}$$

$$\Delta h_{A-B} = 11 \ (0.005) = 0.055 \text{ ft}$$

$$\Delta h_{B-C} = 44 \ (0.02) = 0.88 \text{ ft}$$

$$\Delta h_{A-C} = 0.055 + 0.88 = 0.94 \text{ ft}$$

$$i = \Delta h_{A-C}/L = 0.94 \ (45.5) = 0.021$$

**FIG. 6.14** Data for Darcy's law calculations for drain design. Example 3—interstate highway.

groundwater inflow as estimated with Darcy's law $= kiA = (10$ ft/day)(0.5)(45.5) $= 228$ ft³/day/ft. Then the *total* $Q = 91 + 228 = 319$ ft³/day/ft. And $Q/i = 319/0.021 = 15{,}200$ ft³/day $= kA$, which is the minimum allowable transmissibility of the open-graded drainage layer.

TRIAL 1. Assume an open-graded base drainage layer with a thickness of 9 in. and an effective thickness of 8 in., for $A = 8/12 = 0.67$ ft². Then $k = kA/A = 15{,}200/0.67 = 23{,}000$ ft/day.

TRIAL 2. Assume a base drainage layer with thickness of 6 in. and an effective thickness of 5 in., for $A = 5/12 = 0.42$ ft². Then $k = kA/A = 15{,}200/0.42 = 36{,}000$ ft/day.

TRIAL 3. Assume a base drainage layer with a thickness of 4 in. and an effective thickness of 3 in., for $A = 3/12 = 0.25$ ft². Then $k = kA/A = 15{,}200/0.25 = 61{,}000$ ft/day.

Assume that from a structural standpoint, a 6-in.-thick base drainage layer offers the best choice, then:

CHECK THE REQUIRED $k$ USING FIG. 5.4.

For this example, $W/s = 44.0/0.02 = 2200$. Next, the equivalent infiltration rate of the groundwater inflow of 228 $ft^3$/day = 228/45.5 = 5 $ft^3$/day/$ft^2$ = 5 ft/day = 60 in./day = 2.5 in./hr; then the total equivalent infiltration rate = (rainfall) + (groundwater) = 1.0 in./hour + 2.5 in./hour = 3.5 in./hour. Now, enter Fig. 5.4 for $W/s$ = 2200 and $I$ = 3.5 in./hour, and read the coefficient of transmissibility $C = k_b t_b$ = 180,000 inch-feet/day. Since $t_b$ = 6 in. minus 1 in. = 5 in., the required $k_b$ = 180,000/5 = 36,000 ft/day. This is the same value that was obtained from direct calculations using Darcy's law (Trial 2).

CHECK THE REQUIRED $k$ FOR A DRAINAGE TIME OF 1 HOUR.

Since the collector pipe is located under the shoulder and the width of the drainage layer is a little less than four highway lanes, an approximate check of the drainage time can be obtained by entering the family of curves in Fig. 5.6 for four lanes by this procedure: Follow the line for an effective thickness of 5 in. to its intersection with the horizontal line corresponding to a grade of 0.005, and read a coefficient of permeability of about 11,000 ft/day. This is much less than is required for transmissibility, so it does not control the permeability, and there is no need to make a more precise check of the drainage time-permeability requirement.

RECOMMENDED DESIGN OF OPEN-GRADED DRAINAGE LAYER.

Use a 6-in.-thick open-graded base with a minimum $k$ of 50,000 to 60,000 ft/day to provide a small added factor of safety to allow for possible increases in groundwater flows over the estimated rate; place the open-graded drainage layer on a suitable subbase that will provide filter protection and have sufficient permeability to allow unobstructed flow of groundwater into the drainage layer.

CHECK SIZE OF DRAIN PIPE AND OUTLET PIPE SPACING.

Use the nomograph in Fig. 5.12, and find the following values: longitudinal slope = 0.005, $I$ = 3.5 in./hour, trial pipe diameter of 10 in. (CMP) or 8 in. (smooth bore). Lay a straight-edge between $I$ = 3.5 in./hour and width of pavement $W$ = 44 ft to pivot line 2; connect another straight-edge between the pipe gradient = 0.005 and pipe diameter = 10 in. (CMP), to pivot line 1; then connect by straight-edge between the points on the two pivot lines to read distance between outlets = 250 ft.

**Example 4 Airport Apron (150-ft Half-width).** Assume that this 300-ft-wide apron, with sloping distances of 150 ft to each side of centerline, has a longitudinal grade = 0.002, a cross slope = 0.01 (unbroken slope); design precipitation rate = 1.5 in./hour; design infiltration rate = 0.4 × design precipitation rate = 0.4(1.5) = 0.6 in./hour; no other significant inflows; erodible subgrade; freezing climate; use 2-hour drainage time (see Fig. 6.15). Drain porosity = 0.3.

First, use Darcy's law to analyze flow conditions in a 1-ft-wide strip of pavement and open-graded base drainage layer in the direction of flow. As shown in Fig. 6.15, such a flow path has a length of 158 ft, and the hydraulic gradient is almost exactly equal to the cross slope = 0.01. For the design infiltration rate of 0.6 in./hour, or 1.2 ft/day, the total quantity $Q$ = 1.2(158) = 190 cu ft/day/ft. So, $Q/i$ = 190/0.01 = 19,000 ft³/day = $kA$, the transmissibility needed in the open-graded drainage layer.

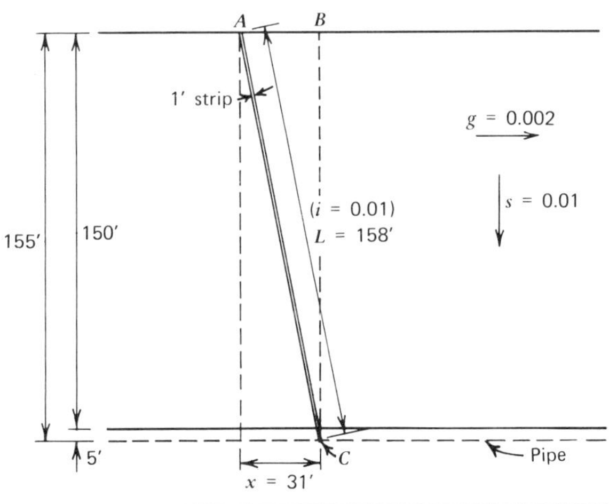


$x = 155'\ (0.002/0.01) = 31$ ft

$L = \sqrt[2]{155^2 + 31^2} = \sqrt[2]{24{,}000 + 960} = \sqrt{24{,}960} = 158$ ft

$\Delta h_{A-B} = 31\ (0.002) = 0.062$ ft

$\Delta h_{B-C} = 155\ (0.01) = 1.55$ ft

$\Delta h_{A-C} = 0.062 + 1.55 = 1.61$ ft

$i = \Delta h_{A-C}/L = 1.61/158 = 0.0102$; use 0.01

**FIG. 6.15** Data for Darcy's law calculations for drain design. Example 4—airport apron.

TRIAL 1. Assume a base drainage layer with a thickness of 8 in. and an effective thickness of 7 in., for $A = 7/12 = 0.58\ \text{ft}^2$. Then $k = kA/A = 19{,}000/0.58 = 33{,}000$ ft/day.

TRIAL 2. Assume a base drainage layer with a thickness of 6 in. and an effective thickness of 5 in., for $A = 5/12 = 0.42\ \text{ft}^2$. Then $k = kA/A = 19{,}000/0.42 = 45{,}000$ ft/day.

Assume that from a structural standpoint there would be no advantage in using a base drainage layer more than 6 in. in thickness, then:

CHECK THE REQUIRED $k$ FOR A DRAINAGE TIME OF 2 HOURS WITH DARCY'S LAW.

If the drainage time is not to exceed 2 hours, water must be able to flow the maximum seepage distance of 158 ft in not over 2 hours or $2/24 = 1/12$ day, and the seepage velocity must be at least $158/(1/12) = 1900$ ft/day. Then, since $v_s = ki/n_e$, the required $k = v_s n_e/i$, and if the effective porosity $n_e = 0.3$, $k = 1900(0.3)/0.01 = 57{,}000$ ft/day. This is greater than was required for transmissibility; hence it is the controlling criterion. (Using the curves in Fig. 5.7, a coefficient of permeability of 48,000 ft/day is indicated for this example; since the hydraulic gradients used in Fig. 5.7 are slightly larger than the slope of the open-graded drainage layer, the permeabilities obtained are less than when the gradient is assumed equal to the slope of the layer, as in these examples.)

RECOMMENDED DESIGN OF OPEN-GRADED DRAINAGE LAYER.

Use a 6-in.-thick layer of open-graded material with a minimum coefficient of permeability of 60,000 to 70,000 ft/day over a subbase that will provide filter protection. Suitable gradations for materials for this open-graded base drainage layer and its subbase would be those given in Fig. 5.9*b*.

CHECK SIZE OF DRAIN PIPE AND OUTLET PIPE SPACING.

Use the nomograph in Fig. 5.12 and find the following values: longitudinal slope of pipe = 0.005 (dug out deeper than grade of pavement to permit smaller pipe to be used), $I = 0.6$ in./hour, and a trial pipe diameter = 8 in. (CMP) or 6 in. (smooth bore). Use straight-edges as in preceding examples to obtain a required outlet spacing of 200 ft. These values for pipe diameters, slope, and outlet spacing appear reasonable; however, the relative costs of other combinations should be checked.

## 6.4 TRANSMISSIBILITY CHART

To relate the discharge needs of the four examples in Section 6.3 with transmissibilities readily obtainable with open-graded bases Fig. 6.16 was prepared. Curves are given in Fig. 6.16 for transmissibilties

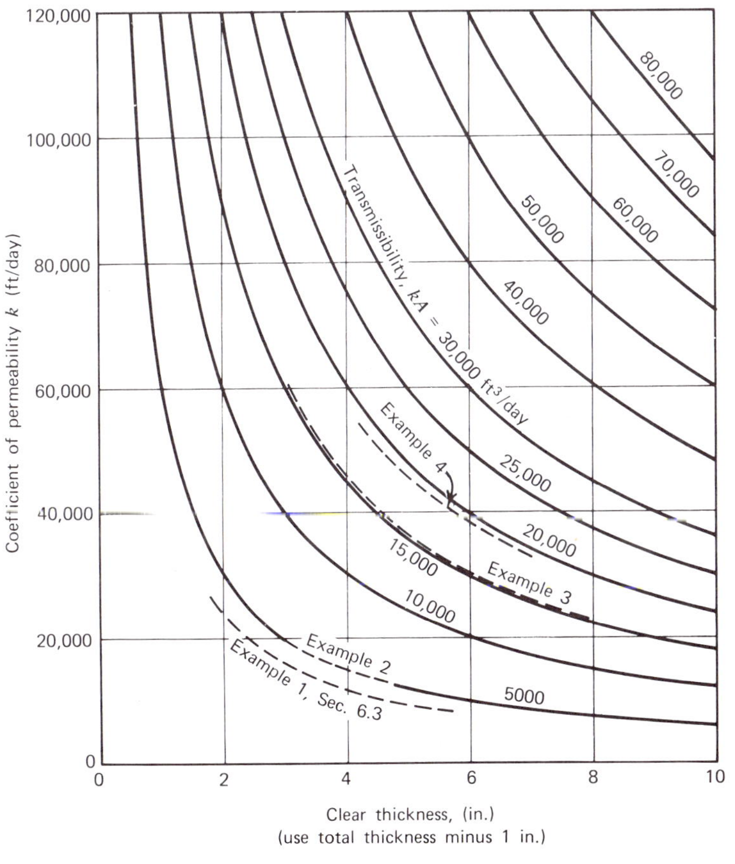


**FIG. 6.16** Combinations of permeability and layer thickness giving transmissibilities from 5000 $ft^3$/day to 80,000 $ft^3$/day (per linear foot of drain).

from 5000 $ft^3$/day to 80,000 $ft^3$/day *per linear foot of drain.* Each curve allows determination of a wide range of combinations of clear thickness of open-graded drainage layer and coefficient of permeability that provide a given transmissibility. By interpolating among the curves Fig. 6.16 can be used for designing drainage systems for any transmissibility within the range of the chart. For example, assume that a transmissibility of 35,000 $ft^3$/day is needed. This can be obtained by using an 8-in. clear thickness of open graded material (total thickness = 8 in. + 1 in. = 9 in.) with a coefficient of permeability of 53,000 ft/day; a 6-in. clear thickness (total = 6 + 1 = 7) with a coefficient of permeability of 70,000 ft/day; a 4-in. clear thickness (total = 4 + 1 = 5) with a permeability of 105,000 ft/day; and so forth.

Clear thicknesses of open-graded drainage layers range from under 2 up to 10 in. in Fig. 6.16, whereas coefficients of permeability vary from under 6000 ft/day (2 cm/sec) to 120,000 ft/day (42 cm/sec).

The relative positions of the four examples within the overall range of Fig. 6.16 are shown by the dashed curves on this chart. It is recalled that the required transmissibilities of these examples are as follows:

$$\begin{aligned} \text{Example 1} &= 4000 \text{ ft}^3/\text{day} \\ \text{Example 2} &= 5000 \text{ ft}^3/\text{day} \\ \text{Example 3} &= 15{,}200 \text{ ft}^3/\text{day} \\ \text{Example 4} &= 19{,}000 \text{ ft}^3/\text{day} \end{aligned}$$

## References

Cedergren, H. R. (1967), *Seepage, Drainage, and Flow Nets,* John Wiley & Sons, Inc., New York, pp. 334–343.

Childers, F. A. (1968), "Drainage Material Study," Georgia State Highway Department, October, 1968.

Federal Highway Administration, FHWA (1973a), *Guidelines for the Design of Subsurface Drainage Systems for Highway Structural Sections,* p. 2 and 6–7; (1973b), *Final Report,* p. 187.

Lovering, W. R., and H. R. Cedergren (1962), "Structural Section Drainage," *Proceedings,* International Conference on the Structural Design of Asphalt Pavements, University of Michigan, Ann Arbor, Michigan, August 20–24, 1962, p. 776.

Strohm, W. E., E. H. Nettles, and C. C. Calhoun (1967), "Study of Drainage Characteristics of Base Course Materials," Highway Research *Record* No. 203, Highway Research Board.

U.S. Army Corps of Engineers (1974), "Methodology and Effectiveness of Drainage Systems for Airfield Pavements," prepared for Construction Engineering Research Laboratory (CERL), Champaign, Illinois, Chicago District, by H. R. Cedergren.

CHAPTER SEVEN

# Safeguarding Subsurface Drainage Systems

A subsurface drainage system that is properly designed and constructed in accordance with carefully prepared plans and specifications should normally function for the intended life of the pavement with relatively little maintenance. Nevertheless, a number of precautions should be taken during the construction and during the life of a system to ensure that it will function as intended. Some of the measures that are considered important are outlined in this chapter.

## 7.1 CARE DURING CONSTRUCTION OF PAVEMENTS AND DRAINAGE SYSTEMS

Some of the phases of road and airfield construction that are important to the proper functioning of subsurface drainage systems are the following.

### Subgrade Preparation

Both the natural foundations and the compacted embankments that make up the subgrades for structural sections should be well consolidated and thoroughly compacted prior to the placement of the structural section. Otherwise differential or uneven settlement is likely to produce sags in the grade that may trap and hold water in local pockets. If such problems are to be kept to a minimum, it is important that subgrades and embankments be compacted to a specified percentage of the "maximum density" by being placed at or near the "optimum moisture content." In the interest of saving a few pennies per cubic yard of compacted materials, moisture requirements are sometimes waived. Frequently, when this is done, noticeable sags can be seen in the pavements within a few years after their completion, and the corrective measures that are eventually required can cost hundreds of times the money saved. To minimize post-construction settlement or consolidation, all unsatisfactory foundation materials should be stripped or other measures taken to properly consolidate compressible formations before pavements are completed.

To further limit the probability of the formation of built-in water traps, grading operations should be true to the lines and grades and tolerances specified. Some engineers recommend that finished subgrade surfaces be graded to the same tolerances as the finished pavement so that the soaking of water into the soil will be held down to the minimum. Whenever possible a slope of at least 1 percent should be used. It should never be less than ½ percent. Completed subgrades should never be used as haul roads by heavily loaded vehicles that can overstress, deform, or rut the subgrade.

Figures 7.1*a* and *b* show sags in pavements that evidently were caused by settlement or consolidation that occurred after construction. The wide pavement in Fig. 7.1*a* evidently had insufficient cross slope to allow for the consolidation that took place under the heavy wheel loads of planes that traversed this area. Insufficient compaction at a culvert evidently was responsible for the settlement that has occurred at the highway location shown in Fig. 7.1*b*.

### Construction of Open-Graded Drainage Layers and Filters, Bases and Subbases

In the construction of a subbase, base, or special "filter" layer under an open-graded drainage layer, care must be taken to avoid any disturbance or contamination of these layers. The use of partially completed highway or airfield pavement structural sections as haul roads can cause irreparable damage; therefore, heavily loaded earth-moving or other construction equipment that can cause any kind of overstress or damage should never be allowed to operate on any layer of a partially completed structural section.

Dirt and other foreign matter must be kept off the tires and tracks of earth-moving and other construction equipment to avoid the dropping of harmful amounts of these materials on any of the layers. Rainfall or other sources of surface water should not be permitted to wash mud and dirt fines onto these layers during the time they are exposed, and measures must be taken to prevent this from happening. All operations should be carried out in a uniform, workmanlike manner, with all surfaces being kept true to grade, without ruts, depressions, humps, or other irregularities.

If concentrated flows from springs are encountered during construction or are expected after completion of a pavement, all such areas

**FIG. 7.1** Settlement and traffic compaction can create sags that impede surface drainage. (*a*) Large "birdbath" on taxiway of an airfield, at side of wide apron; may be caused by traffic consolidation; Great Lakes area; fall, 1972. (Courtesy CERL.[1]) (*b*) Sag in an interstate highway in New England; probably caused by insufficient compaction of backfill at culvert; September 1970.

**FIG. 7.2** Mud has washed on the base course for an airfield pavement under construction; unless removed it can cause problems with the completed pavement; eastern state; fall, 1972.

should be given special treatment to be sure that all inflows will be properly removed (e.g., see Fig. 4.19). Whenever necessary, pipe drains should be installed to remove excess water to allow the work to be properly carried out "in the dry."

The specifications should require that any unsatisfactory material or workmanship be corrected. Figure 7.2 shows mud that has washed onto a base course by rainfall during the construction of an airport runway. If such material is not removed, the quality of the finished work will be impaired.

All aggregates and other materials used in drainage layers and other parts of structural sections should be kept within specification limits for grain size, quality, and so forth. To minimize segregation, filter and other materials should be thoroughly dampened just before being placed. Tests for compliance with gradation specifications (particularly, drainage materials) should be made on samples taken after the materials have been placed and compacted in the project. Any contaminated or segregated materials should be removed and replaced with materials that comply with the specifications.

### Construction of Wearing Courses

After an open-graded base drainage layer has been placed, no subsequent construction activity should be permitted that can impair the integrity of the drainage layer. If tack coats or primes of asphaltic materials are applied in the preparation of surfaces for placement of asphalt concrete wearing courses, excessive amounts of these materials must not be permitted to penetrate the open-graded bases.

Where a PCC wearing course is to be placed over an open-graded base, procedures should be developed that do not allow significant amounts of cement mortar to flow downward into the open-graded base. This is considered very unlikely to be a problem if the slump of the concrete is always under 3 or 4 in. Nevertheless, if very coarse open-graded bases are used, a thin layer of stone chips at least one-fourth the 15 percent size ($D_{15}$) of the open-graded base could be used as a leveling or smoothing course and as a "choker" to aid in keeping mortar out of the drainage layer. Any such courses should be rolled with a three-wheel steel roller before the concrete is poured. If any doubt exists regarding the probability of mortar penetrating into an open-graded base drainage layer, a small test slab should be poured and vibrated to simulate actual construction procedures as closely as possible. When the concrete has set sufficiently, the slab should be lifted off the open-graded base and the base inspected for mortar penetration. Procedures should be modified as necessary to prevent any harmful penetration of mortar into the drainage layer.

### Collector Pipes

Immediately before installation, collector pipes should be inspected to be sure they are undamaged and have a full, clear water passageway. Before pipes are installed in either a shallow V trench in an open-graded base drainage layer or in a rectangular trench (or as otherwise provided in the plans and specifications), the specified bedding material should be placed and properly compacted. Figure 7.3 shows an asphalt coated PMP (perforated metal pipe) on a bedding layer in a trench prior to backfilling the balance of the trench with drainage aggregate.

The material used for filling the balance of a trench should be placed by methods that ensure no harmful segregation. During the placement of the pipes and the materials around and over them, care must be taken to avoid breaking or collapsing the pipes and to prevent the intrusion into the pipes of debris or other foreign matter that could impede the free flow of water. When pipes are being laid the longi-

**FIG. 7.3** Asphalt-coated, corrugated PMP (perforated metal pipe) in a trench; on aggregate bedding; part of new underdrain system for a runway at a major airport August 1972.

tudinal grade of the pipes should be checked frequently to be sure that a uniform positive grade is maintained continuously in the direction of the flow of water. After the pipes have been installed, the continuity of the passageway should be checked by drawing an object through them or, when the alignment permits, by inserting a light and visually inspecting from one point to another farther along the alignment.

In all areas where pipes are provided to collect water from surrounding drainage blanket or trench backfill material, the perforations or slots should be located on the bottom (or all around). In areas where pipes are only conveying water (such as in outlets), the pipes should have no perforations or slots. Where perforations or slots are used, they should always be of such dimensions that the adjacent filter or drainage material cannot work into or clog the openings (see Eqs. 5.10 and 5.11).

### Drain Trench Backfill

After the pipes have been placed in a trench, the backfill material should be placed promptly to minimize any entry of foreign matter, debris, mud, and so forth into the trench. Drainage materials around pipes should be placed with conveyors, spreading boxes, or other equipment that will assure the uniform placement of the material without segregation and without dislodging or damaging any filter cloths placed on the sides and bottom of the trench (see Fig. 6.8). Compaction of the trench backfill material should be as set forth in the specifications and should be sufficient to prevent settlement of overlying pavement or shoulder under the impact of traffic. These materials should not be *overcompacted* because this may reduce permeability below required levels, particularly when graded materials are used. Any segregated or contaminated materials should be removed and replaced with materials that will comply with the specifications.

### Outlets and Outlet Markers

Usually outlet pipes will not be perforated, because their purpose is to convey water from the structural section to suitable disposal points. Normally the backfill around an outlet pipe should be of a low-permeability material that will reduce the possibility of soil piping along the pipe or trench. When suitable impermeable materials are not available, cut-off collars should be located at appropriate intervals along the outlets. To keep rodents and other animals, birds, and so forth from building nests or depositing debris in the pipes, a suitable screen should be provided that can be pushed outward under a small head of water if grass or other matter should collect on the inner side of the screen. In some areas where high flows occur intermittently in the ditches into which outlets discharge, suitable flap gates may be needed to prevent the washing of mud and other matter into the outlets during these high levels. A detail for an outlet that is suggested in the *Final Report* of the FHWA *Guidelines* project (FHWA 1973) is given in Fig. 7.4.

Outlet pipes should be protected and marked with a suitable post as soon as the installation is completed. In most areas a round or square wood post approximately 48 in. long can be installed to a depth of 18 to 24 in. by using a power auger to make a suitable hole. Wood posts

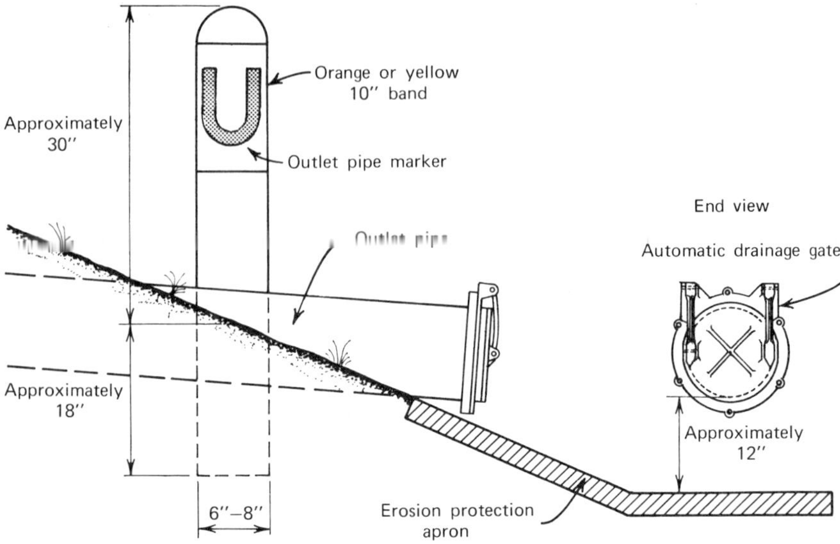


**FIG. 7.4** Detail of an outlet pipe and marker (FHWA, 1973).

should be treated to prevent attack from boring insects and to retard rot. In some cases metal posts may be preferred to wood. A rugged wooden post is less susceptible to displacement and damage than a light metal post; however, some engineers feel that wood posts may constitute a hazard to motorists and prefer lighter metal posts, particularly in areas of deep frost penetration. Metal posts will need more frequent replacement than wood. Posts of wood provide more protection against mowing machines and the like than metal ones do and therefore are generally preferred. Posts with identification markers not only alert maintenance workers to the presence of outlets, but if placed immediately also serve to alert construction forces in general and landscaping contractors and workers in particular to the existence of outlets. Landscaping of highways and airfields often includes the placement of a layer of topsoil over areas of shoulders and slopes that will be turfed for erosion control. The indiscriminate spreading of topsoil has caused unmarked outlets to be covered with several inches of impervious topsoil, blocking the flow of water out of subsurface drains. Figures 7.5*a* and *b* show excellent marker posts.

## 7.2 CARE IN MAINTENANCE OF PAVEMENTS AND DRAINAGE SYSTEMS

The old adage, "an ounce of prevention is worth a pound of cure" applies with full force to pavements of all kinds. The careful, timely maintenance of pavements and their appurtenant facilities is essential to the safeguarding of the capital investments and preserving the quality of all of the facilities. The provision of properly designed and constructed subsurface drainage systems for pavements can sub-

**FIG. 7.5** Location and design of outlet markers is important. (*a*) An excellent drain outlet marker; state highway in south-central state; November 1970. (*b*) Another good drain outlet marker at subdrain discharging above a concrete-lined ditch; south-central state.

**FIG. 7.5** (Continued).

stantially reduce total maintenance work on roadbeds and airfields. Some precautions are needed, however—mainly in the normal maintenance operations—to make sure that the subsurface drainage systems will not be damaged by the careless conduct of any of these operations.

The basic principle to be followed is that nothing should be done by maintenance (or lack of maintenance) that will reduce the efficiency of the drainage systems. Some of the kinds of maintenance operations that are considered of greatest importance are discussed in subsequent paragraphs.

### Surface Cleaning Operations

Although slide debris, mud, blow sand, wind-blown silt, and other matter that gets on pavement surfaces needs to be promptly removed to prevent skidding hazards to the users of the facilities, its removal also helps to prevent large amounts of such materials from working through joints or porous pavement into the supporting bases and drainage layers, and its removal should be encouraged from this standpoint also.

Where pavements are provided with highly permeable, macadam types of base drainage layers, maintenance forces should not allow large amounts of hot or liquid asphalts, emulsions, and so forth to penetrate through joints or porous wearing courses into the open-graded drainage layers, for this may fill up pores in these layers and cause some local obstructions in flow.

No foreign matter should be dumped at outlets because free discharge of water is essential to the preservation of the discharge capabilities of the entire subsurface drainage system.

### Outlet Pipes

Outlet pipes are the only parts of drainage systems that are exposed and subject to direct damage or blockage by lack of proper maintenance, improper maintenance, or landscaping operations. The upkeep of outlets should not require an inordinate amount of time on the part of maintenance forces, but these outlets should be inspected periodically (at least once every 3 months) to be sure they are not blocked by weed or grass growth, bird or animal nests, mud from slope erosion, slide debris, or other matter that can reduce discharges from the pipes. If any foreign matter is found at or in the ends of pipes, it should be promptly removed, and the conditions causing it to accumulate should be corrected because blockage can greatly reduce or nullify the benefits of a drainage system.

Figures 7.6*a* and *b* show outlets that became completely blocked with slide debris and other foreign matter. Such problems are most likely to develop when projects are new and erosion control has not been kept ahead of bad weather conditions. Weed or grass growth that can impede flows from outlets can usually be prevented by the provision of a concrete splash block and sterilization of the soil immediately surrounding an outlet.

### Outlet Markers

Clearly visible outlet pipe markers (a label with the word "underdrain," "drain," "drain outlet," etc. on a protective post) serve as reminders to maintenance forces that outlets are there (see Figs. 7.5 and 7.6). If the markers are visible from the edge of the road or airfield runway, taxiway, and so forth, it facilitates finding them for inspections. Outlet markers are sometimes damaged by maintenance operations and vandalism and therefore should also be inspected at periodic intervals. Plan maps of pavements should clearly show the locations of all drain outlets so that none will be missed by the inspection forces. If

**FIG. 7.6** Outlet pipes must not become clogged or buried. (*a*) Good marker and post, but drain became completely buried by slope wash material; central state; fall, 1970. (*b*) Drain pipe that had been completely covered by mud; interstate highway in Great Lakes area; October 1970.

**FIG. 7.7** Good surface drainage facilities protect airfield and airport pavements. (*a*) A good inlet adjacent to an airfield taxiway; in a semiarid climate; September 1972. (Courtesy CERL.[1]) (*b*) A good inlet in a PCC pavement on an airfield apron; West Coast area; September 1972. (Courtesy CERL.[1])

any are found to be damaged or missing, they should be promptly repaired or replaced.

The two basic objectives of maintenance of outlet markers are (1) to keep the markers plainly visible from the edge of the roadway or other paved area and (2) to make sure that the protective posts are always present at every outlet.

### General Maintenance Operations

All those involved in the general maintenance operations of highways, airfields, and the like should be made aware of the presence of subsurface drainage facilities and their outlets and the need to avoid any activity or usage of materials that can impede the free flow of water from these systems or in any way damage any of their components. When outlets discharge into natural or constructed ditches, periodic checks should be made to be sure that the ditches are kept free of obstructions.

Where inlets or other structures are provided for the collection and removal of surface water from pavements and adjacent areas, these facilities should always be kept free of any kind of debris that can block the flow of water. Any structural damage, undermining of the foundations, or other damage should be promptly repaired. Some well-maintained inlets at major airfields are shown in Figs. 7.7*a* and *b*.

Maintenance forces should be on the alert for any reconstruction activities by the agency operating a pavement facility or by subdividers, land developers, commercial developers, or any others making changes or improvements (outside of the project rights of way as well as within) that can disrupt the equilibrium as it existed at the time the pavements and drainage systems were constructed. If, for example, a natural drainage course or ditch that is depended on for the removal of water from a pavement drainage system should be filled in for future remodeling or enlarging of an airfield, widening of a highway, or similar project, all underdrain outlets should be connected to suitable pipe headers and discharge lines that will assure continued free gravity flow from the drainage system after the alterations have been completed.

Appurtenant structures such as lamp holes, manholes, drop structures, energy dissipation structures, splash blocks, and other facilities should be periodically inspected, and any significant damage or inadequacies should be promptly repaired or corrected.

## References

Federal Highway Administration, FHWA (1973), "Final Report" of the *Project for Development of Guidelines for the Design of Subsurface Drainage Systems for Highway Structural Sections*, Fig. 14, p. 104.

CHAPTER EIGHT

# Economics of Undrained and Drained Pavements

## 8.1 INTRODUCTORY COMMENTS

The true economics of any engineering construction work depends on the total ultimate cost over the useful life of the facility, including not only the original expenditures but all significant maintenance and replacement costs. The value returned by the investment can be measured in the amount of trouble-free service provided over the useful life of the facility. If cost-effectiveness trade-off studies are based on original costs only, without regard to either the actual useful service life of a facility or the upkeep and replacement outlays, these studies can be quite meaningless.

Good drainage offers an inexpensive, although seldom-used method for getting the most for the money spent on pavements. Finding well-drained pavements these days is very difficult, however, because the predominating philosophy of pavement designers and builders is that subsurface drains are an "expensive luxury." This belief persists in spite of the fact that large outlays are being required to repair and replace severely damaged pavements—often long before the design life is reached—and at costs far exceeding any that would have been required for drains. An indication of the magnitude of the problem is given in a GAO (General Accounting Office) report to the Congress, summarized in *Civil Engineering-ASCE* (1970). According to that article, the GAO report in early July 1970 says that the problem of overlayment of relatively new interstate highways is a major one, and costs will run into "hundreds of millions" of dollars well before any of the roads have reached their 20-year design age.

These accelerating maintenance and repair costs are due to a number of causes, including increasing volume of heavy truck traffic, but there can be little doubt that the damaging effects of heavy wheel loads on roadbeds that are in a *supersaturated* or *flooded state* is contributing heavily to these unexpected costs.

In a report summarizing a comprehensive investigation of field performance of drainage and erosion control facilities at 54 airfields and airports in the United States in the period from October 1966 to June 1972, Grace et al. (1973) concluded that, "Experienced qualified personnel agree that the initial installation of adequate facilities provides satisfactory performance and requires least maintenance which in the long run provides greater economy than that realized by multiphased stage improvements and remedial measures attempted later. . . ." There is a good deal of evidence supporting this conclusion; many airfields and airports throughout our country are experiencing major maintenance and replacement costs for undrained pavements damaged by heavy traffic. For example, over $7 million was spent for the complete reconstruction of a failed runway-taxiway system at a major municipal airport in the United States. According to reports by the engineers who designed the new pavements, the old pavements deteriorated very rapidly under heavy traffic of very large planes. Not only was there extensive spalling and cracking, but there was evidence of loss of support by the saturated foundation soil. Because of large amounts of down time and mounting repair costs, it was decided that the city would provide a landing surface that would be not only completely safe but free of broken pieces of pavement and other debris that could ruin jet engines if drawn into them.

The designers also reported that the old runway, which had been built in the 1940s with a 9-in.-thick PCC wearing course on a 6-in.-thick aggregate base over the clay subgrade, had been provided with no subsurface drainage system at all. Poor drainage with the entrapment of water in the structural section and on the subgrade appeared to be the major cause of the failure. These engineers decided to design a new runway-taxiway system with a complete underdrain system. The added cost of the drainage system, with over 15 miles of pipe drains under a special drainage layer, was less than 5 percent of the total cost.

After riding on this pavement system in 1972 and examining parts of it in detail, the author concluded that the drainage system will pay for itself many times over.

When major repairs or total reconstruction of runways and other pavements are required at important airports or airfields, the losses to the agencies or airlines and to the passengers from delays in landing and taking off can greatly exceed the actual costs of the repairs or reconstruction. At one major airport it was estimated that losses to the airlines and passengers averaged more than $200,000 for each day of congested traffic causing ½- to 1-hour delays in landings and takeoffs for large numbers of aircraft and large numbers of passengers. For a month, this would represent a loss of at least $6 million.

In surveys the author has participated in for federal agencies to investigate the possible effects of excess water on rates of pavement damage, many pavements were found that had been severely damaged by the combined effects of heavy traffic and excess water in structural sections. Assessing the magnitude of the losses was difficult, because most major maintenance and replacement costs are paid out of "construction" funds rather than from "maintenance" funds. Thick overlays, grinding, and other major repairs and replacements that are beyond "normal" maintenance generally do not show up under maintenance budgets. Therefore the costs of improving undrained pavements are not always readily available, but, the benefits of building truly well drained pavements have been outstanding. Several examples of fine experiences with good drainage are described in Sec. 9.5.

The experience records of modern pavements support the concept that any well drained highway or airfield pavement is inherently superior to the same pavement, poorly drained. With rare exceptions it has been difficult to find truly well drained pavements. A notable example, described in Sec. 9.5, is a short section of a county road on Kneeland Hill in Humboldt County, California. Located in a very high rainfall area (40 to 80 in./year) and with large amounts of sidehill seepage, this stretch of road was reconstructed in 1967 after two prior failures caused by excess water. The prior reconstructions (Cedergren and Lovering, 1968) had been made with types of base and drainage materials in general usage. Although the volume of traffic is not large, heavily loaded logging trucks use the road. Even though the previously existing pavement had been reconstructed on a layer of so-called "permeable material," it failed within a short time from excess water, as evidenced by the flow of water out of cracks in the deteriorating pavement.

In the final reconstruction of this section of road by the Humboldt County Public Works Department in 1967, a highly permeable macadam type of drainage layer was constructed under the pavement and base courses. An article in *Rural and Urban Roads* (1972) told about tests that were made on this short stretch of road as part of a study for the FHWA on drainage practices for highways. This Humboldt County section of road was selected for one of the case studies for the FHWA Guidelines project referred to in a number of places in this book. Investigations for that study showed that in spite of the use of this road

**FIG. 8.1** Humboldt County experimental road, rebuilt on highly permeable, macadam-type drain layer after two previous failures from excess water with standard road-building materials; April 1971. In excellent condition in 1974.

by heavy logging trucks (Fig. 8.1) and the two prior failures the reconstruction was in perfect condition after several wet winters. The article goes on to say

> When we drilled into the base and subbase layers, we found an almost unbelievably well drained condition in all layers. The condition of the supporting layers was as different as day from night when compared with other case-study sites [state highway sites], which were all having troubles from excess water. This is a remarkable example of the benefits of good drainage. The only regret is that almost no modern roads are being built with this kind of protection.

Many examples could be cited of pavements which have required very costly repairs because of damage from exposure to free water or *a lack of good drainage.* In 1972 the author inspected a runway of a major airfield (Great Lakes area) that after a number of years of traffic and exposure to the elements had suffered extensive joint spalling and D cracking that had resulted in repairs totaling nearly $2.5 million, and the problems were only temporarily alleviated. Engineers in charge of the maintenance and operation of this airfield said they believed that the major amount of this damage could be blamed on poor drainage and extensive exposure of the runway pavement (PCC) to internal saturation and flooding from the rainfall that soaked into the pavements. The D cracking, spalling, and blow-ups (which are all aggravated by exposure to free water) are adding substantial amounts to our country's repair bills for pavements.

A number of airport and highway engineers interviewed by the author (and other members of interview teams) during site inspections have expressed the opinion that rapid drainage of free water out of pavements could drastically cut the amount of damage from D cracking. When severely D-cracked and spalled PCC pavements reach the state where overlays are needed to prevent hazards from loose material on pavements (FOD), extensive diamond-sawing and chipping of spalled joints are often required to remove deteriorated material (Fig. 8.2). This kind of repair work is extremely costly.

When pavements become severely damaged from heavy wheel impacts on sections filled with water or by D cracking or blow-ups, the loosened material can be extremely hazardous to the users of the pavements. The dangers of loosened material (FOD) to jet aircraft are of

**FIG. 8.2** Deteriorated joints in airfield apron have been extensively diamond-sawed and chipped in preparation for AC overlay; Great Lakes area; August 1972. (Courtesy CERL.[1])

major concern to those operating such aircraft. Flying pieces of aggregate or pavement can also cause hazards to highway users, and large jagged pieces of pavement can cause tire blowouts. In addition, the costs of repairing these damages can be of major economic significance.

The hazards to the users of damaged pavements, as well as the inconveniences and delays, represent major "hidden" costs of the poorly drained pavements. Several aspects of the relative economics of both well-drained and poorly drained pavements are covered in this chapter. The cost comparisons do not include any hidden costs. Nevertheless, they indicate that the well-drained pavements have annual costs about one third less than the undrained pavements.

If major outlays are needed to completely remove and replace severely damaged pavements, such as the runway-taxiway system noted in preceding paragraphs, or extensive overlays, patching, grinding, repairing, and the like are needed, the total long-term cost of the pavement can greatly exceed the initial cost. When the damaging effects of excess water are finally added up, cost analysts can then determine the amount of extra funds that could have been spent originally for drainage systems (or other measures) that would have eliminated the need for such outlays.

In succeeding parts of this chapter some comparisons are made between the cost of undrained and drained designs for pavements on the basis that the useful or effective life of a pavement is reduced in proportion to the "severity factor" of wet impacts and the amount of time excess water is present in the structural section as depicted by curves in Fig. 2.13. Those responsible for making the choice between alternate designs should develop realistic cost comparisons using local unit prices and factors such as anticipated inflation and interest value of the money. The illustrative examples given in this chapter indicate that if severe water damage occurs in an undrained pavement, the long-term cost will be much greater than if good drainage had been provided *at the time of the original construction.*

## 8.2 COST PER SQUARE YARD PER YEAR OF USEFUL LIFE

Comparisons of the costs of alternate pavement designs (such as conventional undrained sections versus well drained sections) should be on the basis of the original costs plus all expenditures for important upkeep, repair, and replacement, amortized over the useful life of the pavements. Original cost alone, with normal maintenance, is not very meaningful if the actual effective service life of a pavement is not taken into consideration. The direct benefits of pavement-drainage systems are the amounts of usable pavements provided and the number of load applications or years of useful service given by the pavements. In this discussion the relative costs of drained and undrained pavements are compared on the basis of unit costs of the amount of area provided [cost/yd$^2$/year].

In Sec. 2.4, "Excess Water and Shortened Pavement Life," it is shown that a pavement which is forced to carry heavy traffic loads while containing excess water can have its useful life greatly shortened. Figure 2.13 shows that if a pavement were designed without considering the actual effects of wet wheel impacts, its life could be substantially reduced. For example, if without damage from excess water a given pavement would have a useful life of 20 years, its life would be reduced to 12 years if excess water were present 8 percent of the time while the severity factor was 10. Cost-effectiveness comparisons in succeeding paragraphs are based on the assumption that pavements have a theoretical design life of 20 years but an actual effective life of only 12 years. They do not include allowances for interest value of capital outlays or price increases because of inflation. As used in the examples, the terms "effective life," "useful life," or "life expectancy" are taken to mean the number of years of service before major overlays or expensive reconstructions are required for significant portions of a given highway or airfield pavement.

**Example 1 Interstate Highway Pavement (PCC on CTB).** In developing this comparison of a conventional undrained highway pavement with a well-drained pavement, the following unit prices are assumed to be reasonable in the continental United States in 1974:

PCC = \$25.00/yd$^3$
Cement-treated base = 15.50/yd$^3$
Aggregate subbase = 6.50/yd$^3$
AC open-graded base drain = 18.50/yd$^3$

The following designs are assumed to have equivalent structural value:

**Conventional Design (effective life = 12 years):**

Concrete pavement = 9 in.
Cement treated base = 6 in.
Aggregate subbase = 6 in.

**Open-graded Base Design (effective life = 20 years):**

Concrete pavement = 9 in.
AC open-graded base drain = 5 in.
Aggregate subbase = 6 in.

(Collector pipes in drains along lower edge; outlet pipes every 200 ft.)

**Analysis of Conventional Design:**

Quantities per yd² (two 12-ft lanes):

$$\text{PCC pavement} = \frac{9 \times 0.75}{27} = 0.25 \text{ yd}^3$$

$$\text{CTB} = \frac{9 \times 0.50}{27} = 0.17 \text{ yd}^3$$

$$\text{Aggregate subbase} = \frac{9 \times 0.50}{27} = 0.17 \text{ yd}^3$$

Cost per yd² (two 12-ft lanes):

| | | | |
|---|---|---|---|
| PCC pavement | = 0.25($25.00) | = | $6.25 |
| CTB | = 0.17($15.50) | = | 2.64 |
| Aggregate subbase | = 0.17($6.50) | = | 1.11 |
| Total | | | $10.00 |

For a useful life of 12 years, the cost/yd²/year will be $10.00/12 = $0.83, thus:

Annual cost per yd² = $0.83

**Analysis of Design with Open-graded Base:**

Quantities per yd² (two 12-ft lanes):

$$\text{PCC pavement} = \frac{9 \times 0.75}{27} = 0.25 \text{ yd}^3$$

$$\text{Open-graded base} = \frac{9 \times 0.42}{27} = 0.14 \text{ yd}^3$$

$$\text{Aggregate subbase} = \frac{9 \times 0.50}{27} = 0.17 \text{ yd}^3$$

(Collector pipes and 20-ft long outlets every 200 ft)

Cost per yd² (two 12-ft lanes):

| | | |
|---|---|---|
| PCC pavement | = 0.25($25.00) | $6.25 |
| Open-graded base | = 0.14($18.50) | 2.59 |
| Aggregate subbase | = 0.17($6.50) | 1.11 |
| Pipe drains @$3.50/ft | = 1.1($3.50) (9/24) = | 1.45 |
| Total | | $11.40 |

For a useful life of 20 years, the cost/yd²/year will be \$11.40/20 = \$0.57; thus:

annual cost per yd² = \$0.57

For this example the well-drained pavement design will have an annual cost of (\$0.57/\$0.83) = 69 percent of that of the conventional undrained pavement. Similar comparisons can be made for any designs and any life expectancies for either well-drained or undrained pavement designs.

**Example 2 Interstate Highway Pavement (AC on Aggregate base).** In making this cost analysis, the following unit prices were used as being reasonably representative of 1974 prices in the continental United States:

| | | | | |
|---|---|---|---|---|
| Asphalt concrete | = | \$13.50/ton | = | \$27.00/yd³ |
| AC open-graded base | = | 11.00/ton | = | 18.50/yd³ |
| Aggregate base | | | = | 8.00/yd³ |
| Aggregate subbase | | | = | 6.50/yd³ |

The following cross sections are assumed to have equivalent structural strengths (not necessarily the same as for Example 1):

**Conventional Design (effective life = 12 years):**

Asphalt concrete = 5 in.
Aggregate base = 8 in.
Aggregate subbase = 12 in.

**Open-graded Base Design (effective life = 20 years):**

Asphalt concrete = 5 in.
Open-graded base drain = 4 in.
Aggregate subbase = 12 in.

(Collector pipes along lower edge; outlet pipes every 300 ft)

**Analysis of Conventional Design:**

Quantities per yd² (two 12-ft lanes):

$$\text{AC pavement} = \frac{9 \times 0.42}{27} = 0.14 \text{ yd}^3$$

$$\text{Aggregate base} = \frac{9 \times 0.67}{27} = 0.22 \text{ yd}^3$$

$$\text{Aggregate subbase} = \frac{9 \times 1.0}{27} = 0.33 \text{ yd}^3$$

Cost per $yd^2$ (two 12-ft lanes):

| | | | | |
|---|---|---|---|---|
| AC pavement | = | 0.14($27.00) | = | $3.78 |
| Aggregate base | = | 0.22($8.00) | = | 1.76 |
| Aggregate subbase | = | 0.33($6.50) | = | 2.15 |
| Total | | | | $7.69 |

This represents an annual cost per square yard of $7.69/12 = $0.64; thus:

Annual cost per $yd^2$ = $0.64

**Analysis of Design with Open-graded Base Drain:**

Quantities per $yd^2$ (two 12-ft lanes):

$$\text{AC pavement} = \frac{9 \times 0.42}{27} = 0.14 \text{ yd}^3$$

$$\text{Open-graded base} = \frac{9 \times 0.33}{27} = 0.11 \text{ yd}^3$$

$$\text{Aggregate subbase} = \frac{9 \times 1.00}{27} = 0.33 \text{ yd}^3$$

Cost per $yd^2$ (two 12-ft lanes):

| | | | | |
|---|---|---|---|---|
| AC pavement | = | 0.14($27.00) | = | $3.78 |
| Open-graded base | = | 0.11($18.50) | = | 2.04 |
| Aggregate subbase | = | 0.33($6.50) | = | 2.17 |
| Pipe drains @ $3.50/ft | = | 1.07($3.50) (9/24) | = | 1.40 |
| Total | | | | $9.39 |

This represents an annual cost per square yard of $9.39/20 = $0.47; thus:

Annual cost per $yd^2$ = $0.47

For this example, the well-drained pavement will have an annual cost of ($0.47/$0.64) = 73 percent of that of the conventional undrained section.

**Example 3 Airport Runway Pavement—PCC on Aggregate Base.** To make this analysis, the following unit prices are assumed to be reasonable for 1974 in the continental United States:

PCC (16-in. thick) = \$23.00/yd³
Aggregate base = 8.00/yd³
AC open-graded base drain = 17.00/yd³
Aggregate subbase = 6.50/yd³

In comparing the costs of a conventional undrained section with a well-drained section (with open-graded base drainage layer), the following alternate sections for 10,000 × 300 ft runways with 75-ft-wide taxiways are assumed to provide comparable structural strength:

**Standard Design:**

16-in. PCC
8-in. aggregate base
8-in. aggregate subbase

**Design with Open-graded Base Drainage Layer:**

16-in. PCC
6-in. open-graded AC base drain
10-in. aggregate subbase

Using the foregoing unit costs and design thicknesses, the relative costs of the alternate designs (exclusive of excavation, grading, shoulders, lighting, and other common costs) would be as follows, if it is assumed that the "standard design" (without drainage) has an effective life of 12 years, whereas the "design with open-graded base drainage layer" has an effective life of 20 years:

**Analysis of Standard Design:**

Quantities of materials per yd²:

$$\text{PCC pavement} = \frac{9 \times 1.33}{27} = 0.44 \text{ yd}^3$$

$$\text{Aggregate base} = \frac{9 \times 0.67}{27} = 0.22 \text{ yd}^3$$

$$\text{Aggregate subbase} = \frac{9 \times 0.67}{27} = 0.22 \text{ yd}^3$$

Cost per yd$^2$:

| | | |
|---|---|---|
| PCC pavement | = 0.44($23.00) | = $10.12 |
| Aggregate base | = 0.22($8.00) | = 1.76 |
| Aggregate subbase | = 0.22($6.50) | = 1.43 |
| Total | | $13.31 |

This represents an annual cost per square yard of $13.31/12 = $1.11; thus:

Annual cost per yd$^2$ = $1.11

**Analysis of Drained Design:**

Quantities of materials per yd$^2$:

$$\text{PCC pavement} = \frac{9 \times 1.33}{27} = 0.44 \text{ yd}^3$$

$$\text{Open-graded base} = \frac{9 \times 0.50}{27} = 0.17 \text{ yd}^3$$

$$\text{Aggregate subbase} = \frac{9 \times 0.83}{27} = 0.28 \text{ yd}^3$$

Drains with pipes* = 0.107 ft/yd$^2$

Cost per yd$^2$:

| | | |
|---|---|---|
| PCC pavement | = 0.44($23.00) | = $10.12 |
| Open-graded base | = 0.17($17.00) | = 2.89 |
| Aggregate subbase | = 0.28($6.50) | = 1.82 |
| Drains @ $4.50/ft | = 0.107($4.50) | = 0.48 |
| Total | | $15.31 |

This represents an annual cost per square yard of $15.31/20 = $0.77; thus:

Annual cost per yd$^2$ = $0.77

For this example the well-drained pavement will have an annual cost (for a life of 20 years) of $0.77/$1.11 = 69 percent of the cost of the undrained section. In view of recent experiences with some undrained pave-

* Assumes 53,500 linear feet of underdrains and outlets (outlets @ 400 ft) for 500,000 square yds of pavements = 0.107 ft/yd$^2$.

ments that have had very high repair costs or have had to be replaced before their design age, it is believed that this comparison is not unrealistic. In fact, when all important costs are considered, many conventional undrained pavements will be more than twice as costly as well-drained pavements.

## 8.3 COST PER CYCLE OF TRAFFIC OPERATION

Another useful basis for comparing the cost-effectiveness of undrained and drained pavements is on the cost per cycle of operation of the vehicles using a facility. Examples are given here for both highway and airfield pavements. With respect to highways, the only traffic that produces measurable damage to pavements is that of heavy vehicles, and repetitions of light vehicles (passenger cars, etc.) are not considered in design formulas.

**Example A Highway Pavement** For this example, the unit costs per square yard for the highway pavement in Example 1, Sec. 8.2, will be used as follows: Undrained design, 9-in. PCC on 6-in. CTB and 6-in. aggregate subbase for a cost of $10.00/yd$^2$ or $13.30/ft of 12-ft lane. Drained design, 9-in. PCC on 5-in. open-graded base and 6-in. aggregate subbase and pipes for a cost of $11.40/yd$^2$ or $15.20/ft of 12-ft pavement lane.

Assume that the undrained pavement design is damaged at a faster rate than allowed for in the design chart or formula. Then, assuming 1000 18,000-lb axle loads/day/lane, the cost per axle load per lane mile would be as follows for an effective life of 12 or 15 years:

Effective life of 15 years: 1.3c/18,000-lb axle load

Effective life of 12 years: 1.6c/18,000-lb axle load

If the drained design will have a useful life of 20 years (or longer), the cost per 18,000-lb axle load per lane mile would be as follows:

Effective life of 20 years: 1.1c/18,000-lb axle load

Effective life of 25 years: 0.9c/18,000-lb axle load

This basis of comparison shows that the drained design has the potential for saving appreciable costs over the life of a major pavement system.

**Example B Airfield Pavement** For this example, use the unit costs per square yard for the airport runway pavement in Example 3, Sec. 8.2, as follows: 500,000 yd$^2$ of pavements; standard undrained design, 16-in. PCC on 8-in. aggregate base and 8-in. aggregate subbase @ $13.31/yd$^2$. Drained design, 16-in. PCC on 6-in. open-graded AC drainage layer and 10-in. aggregate subbase @ $15.31/yd$^2$.

**TABLE 8.1** Cost/Cycle of Heavy Plane Operations on 500,000 yd² of PCC Pavements, Undrained and Drained[a]

| Effective life | | Cost/cycle of operation | |
|---|---|---|---|
| Cycles | Years | Undrained | Drained |
| 60,000 | 12 | $111 | — |
| 75,000 | 15 | $89 | — |
| 100,000 | 20 | — | $77 |
| 125,000 | 25 | — | $61 |

[a] Assumes that increased severity factors of undrained pavements reduces effective life; early 1974 prices; costs per cycle rounded to nearest dollar.

The actual "effective life" of the undrained design will be based on the assumption that 8 percent of the wheel impacts that occur with excess water produce 10 times the damage of each operation when no free water is present. This would result in an "effective life" of 60 percent of that of the drained design, or (0.60) (20) = 12 years.

For a 12-year effective life during which (0.6) (100,000) or 60,000 cycles of operations occur, the cost/cycle would be as follows:

Cost of undrained pavements = 500,000 ($13.31) = $6,655,000

Cost/cycle = $6,655,000/60,000 = $110.92

For the drained design with an effective life of 20 years during which 100,000 cycles of operations occur the costs would be:

Cost of drained pavements = 500,000 ($15.31) = $7,655,000

Cost/cycle = $6,655,000/60,000 = $110.92

If the drained design actually lasted 25 years because of its effective drainage system, the cost per cycle for 25 years would be as follows:

Cost of drained pavements = $7,655,000

Cost/cycle = $7,655,000/125,000 = $61.24

The costs per cycle for a range of effective life cycles of the undrained and drained designs used in this example are summarized in Table 8.1.

## 8.4 THE ECONOMICS OF AGGREGATES FOR DRAINAGE

When aggregates are used for the drainage of civil engineering works such as dams, spillways, reservoir linings, and pavements, one of their primary functions is the *removal* of water. Where aggregates have to carry water appreciable distances, they are serving as *conveyors* or *conductors,* just as the pipes in water supply systems or sewerage systems are conveyors. In the past little attention has generally been given to the actual *needs* of drains with respect to their water-removing or conducting capabilities, with the result that great deficiencies have often existed. The rational seepage-analysis methods outlined in other chapters show how to design drains that will have the necessary discharge capacities to do what is expected of them. If these methods are applied, surprising information is obtained about the economics of aggregates that are used for drainage purposes.

When the discharge capabilities of drains has not been determined, selection of aggregates for drainage purposes has generally led to the use of types of aggregate that are extremely uneconomical as conveyors of water. For example, an earth dam 100 ft high was designed and constructed with an internal drain made of washed concrete sand with a coefficient of permeability of 1 to 2 ft/day ($3.5$ to $7.0 \times 10^{-4}$ cm/sec). Because of the low permeability of the drainage aggregate specified and used, the discharge capacity of the entire drainage system (which cost over $50,000) was only 3 gpm. The cost of this drainage system for removing seepage (its primary purpose) was more than $15,000 for each gpm of discharge capacity (Cedergren, 1967), and serious seepage problems developed from a lack of effective drainage. Another much larger dam that developed seepage problems from a lack of drainage had a multilayered drain costing approximately $1.4 million. This drain had a discharge capacity of about 5 gpm for a cost of $300,000 for each gallon per minute of discharge capacity. Still another example of small return for capital invested was the drainage system for a major freeway constructed in wet, hilly terrain with a 12-in.-thick blanket of so-called "permeable material" to protect the roadbed from water damage. Because the well-graded blend of sand and gravel that was expected to drain this roadbed had a very low permeability (1 to 3 ft/day) this drain cost more than $50,000 for each gallon per minute it was capable of removing. A few years after completion of this highway, heavy repair costs had been incurred and excess water was bleeding out of the deteriorating pavements, whereas the outlet pipes were dry.

Any conveyor or conductor of a material can be rated in terms of the cost of moving a given unit quantity of the material over a given distance. Thus in earthwork it is customary to use the term "station-yard," and in freight-hauling the term "ton-mile" has been used. Cedergren and Lovering (1968) use 1 gpm conducted 100 ft as the unit in making a comparison of the relative costs of various drainage aggregates serving as conveyors of water. Estimates of the quantities of drainage aggregates needed to conduct a given quantity of water some distance can be made (Cedergren, 1974) with Darcy's law,

$$q = kiA$$

Multiplying the right-hand side of this equation by unity ($L/L$), as follows, does not change the identity:

$$q = kiA \frac{L}{L}$$

Then since $AL$ represents a volume of aggregate $V$,

$$q = \frac{kiV}{L} \tag{8.1}$$

In Eq. 8.1, $V$ is the volume of drainage aggregate having a cross-sectional area $A$ and a length $L$. It can then be seen that

$$V = \frac{qL}{ki} \tag{8.2}$$

Thus Eq. 8.2 relates the volume of drainage aggregate needed to convey seepage quantity $q$ a distance $L$ through cross-sectional area $A$ under hydraulic gradient $i$.

Assuming that a drainage layer is flowing full, that the effective gradient is an average hydraulic gradient $i$, and that aggregates cost \$8.00/yd$^3$ (in 1974), the charts in Fig. 8.3 were obtained. If it is recognized that such determinations are somewhat approximate because of the simplifying assumptions, considerable useful information is still obtained. By examining Eq. 8.2 it is apparent that for a given value of $qL$ the amount of aggregate needed (and therefore the cost of the aggregate needed) varies inversely with $k$ and $i$. This is also quite apparent in Fig. 8.3. For a hydraulic gradient of 0.02 (a value not often exceeded in pavement drainage layers), it would cost nearly \$250,000 to convey 1 gpm a distance of 100 ft using a sandy gravel with a coefficient of permeability of 1 ft/day ($3.5 \times 10^{-4}$ cm/sec); whereas the cost would be only \$2.30 to accomplish this amount of drainage using an open-graded aggregate with a coefficient of permeability of 100,000 ft/day (35 cm/sec).

For unit aggregate costs either larger or smaller than \$8.00/yd$^3$, the ordinates in Fig. 8.3 should be multiplied by the ratio of the actual cost to the assumed \$8.00/yd$^3$. Thus if aggregates cost \$12.00/yd$^3$, the ordinates would be multiplied by (\$12.00/\$8.00) = 1.50.

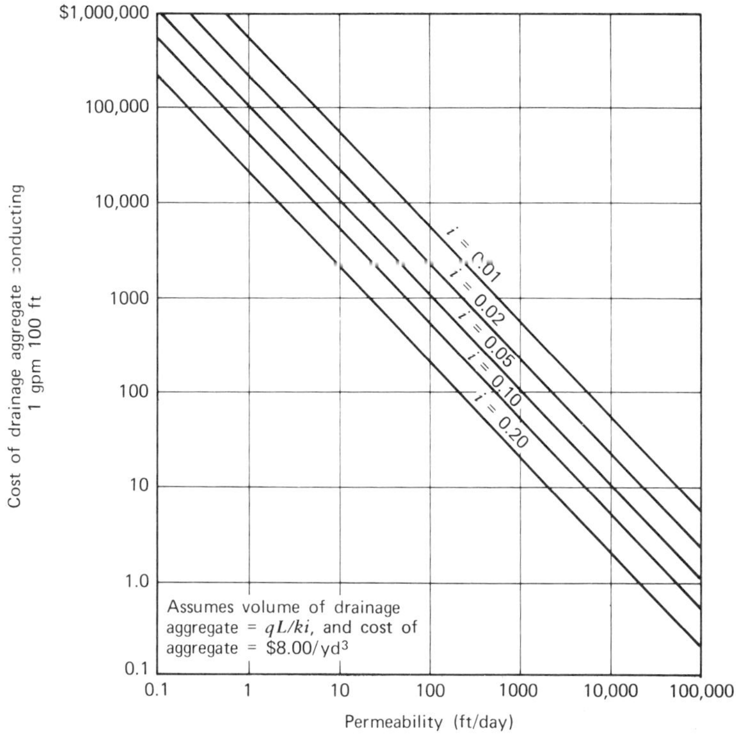


**FIG. 8.3** Cost of drainage aggregates needed to conduct 1 gpm a distance of 100 ft.

The charts in Fig. 8.3 indicate that when aggregate drainage layers are looked on as conveyors of a material (water), highly permeable open-graded aggregates cost far less than conventional blends of sand and gravel. Comparisons of this kind are felt to be valuable aids in developing an appreciation for the inherent capability or *incapability* of various classes of aggregates to serve their basic purpose of rapidly removing free water from pavement structural sections.

## 8.5 COST-TRADEOFF STUDIES OF CORRECTIVE TREATMENTS

In preceding paragraphs the relative costs of undrained and drained pavements are compared on the basis of unit cost per square yard per year of useful life and on the basis of traffic operations over the useful life. These comparisons are somewhat hypothetical, because actual pavements will be kept in operating condition regardless of the costs of repairs or replacements. Another approach to the cost-tradeoff of undrained and drained pavements is on the basis of actual outlays for unusual maintenance such as heavy overlays or digging out and replacing broken sections of important pavements. When cost estimates are made for both a standard undrained design and a drained design, the initial cost of the drained design may be slightly higher than for the undrained design. If an undrained design is selected and deterioration sets in before the design age, however, the undrained pavement will cost the same as the drained pavement when its repair costs equal any extra initial costs for drainage. An illustration of this, using the estimated costs in Example 1, Sec. 8.2, assumes the initial costs of undrained and drained designs are as follows:

Undrained design (9-in. PCC on 6-in. CTB and 6-in aggregate subbase) = $10.00/yd^2$

Drained design (9-in. PCC on 5-in. open-graded base drainage layer of AC and 6-in. subbase) = $11.40/yd^2$

For this example the "break-even" point is reached when repairs amount to $11.40 − $10.00 = $1.40/yd^2$. For repairs over $1.40/yd^2$ (total pavement area), it would be less expensive to have built the drained section.

### Asphalt Concrete Overlays

Assuming that AC overlays cost $0.75 per square yard per inch of thickness (1974), the "break-even" point for overlays alone (using the costs in Example 1, Sec. 8.2) would be reached when the total cost of AC overlays is $1.40/yd^2$ of pavement, or ($1.40/$0.75) = 1.87 in. of overlays have been placed. Any greater thickness of overlays will result in total costs greater than those of the drained design (original construction). This comparison is based on only the construction and materials costs and does not make any allowance for indirect costs, such as the inconvenience to the motoring public and traffic control. An illustration of an overlay being placed on an interstate highway is given by the photograph in Fig. 8.4. Here a thick AC overlay is being laid on an 8-year-old AC pavement that has started to crack up prematurely under the effects of traffic and surface water infiltration.

One of the common problems with overlays (either AC or PCC) is that cracks in the underlying pavement often progress through the overlays and show up as "reflection" cracks. This allows surface water to enter and aggravate any instability problems or other water-induced damage. Figure 8.5*a* shows such cracks that have come through a 7-in.-thick overlay on an airfield apron pavement.

Covering deteriorating pavements with overlays may slow down the rate of damage (if placed before serious deterioration has occurred),

**FIG. 8.4** Thick AC overlay being placed on 8-year-old interstate highway that is cracking prematurely from excess water; south-central state; no groundwater; November 1970.

but in many cases the benefits are of limited value. Figure 8.5*b* shows water and fines being pumped through a 2-in.-thick AC overlay on an interstate highway pavement. This overlay has not stopped the bleeding and pumping that were going on prior to its placement. Figure 8.5*c* shows stains at a crack that has come through a 4-in.-thick AC overlay on an airfield warm-up apron in which water and other matter have been bleeding out during and after rainfalls.

Another problem with overlays is that of obtaining a good bond with the underlying pavement surface. Unless a nearly perfect bond is obtained, pieces of overlay can become separated from the original pavement, producing a very rough, uneven surface and creating hazards from the loosened debris. This is particularly frequent with thin overlays, although it can also occur with relatively thick overlays placed over badly deteriorated pavements.

Although overlays are widely used for reinforcing pavements that are showing distress under traffic, they are no panacea. Also, as previously demonstrated, their costs can soon exceed those of providing good drainage systems in the initial construction.

**FIG. 8.5** Overlays are no panacea for pavement troubles. (*a*) Reflection cracks have come through 7-in.-thick AC overlay on PCC pavement on airfield apron; Great Lakes area; August 1972. (Courtesy CERL.[1]) (*b*) Water and fines being pumped out of 2-in. AC overlay on 12-year-old interstate highway damaged by traffic and excess water; West Coast area; December 1970. (*c*) Stains show bleeding and loss of fines out of airfield warm-up apron; 4-in. AC overlay on 25-in. PCC; Great Lakes area; September 1972. (Courtesy CERL.[1])

(c)

### Grinding of PCC Pavements to Reduce Faulting

Grinding with diamond saws or rotating chopping bits or drills is often resorted to for reducing the step-off or faulting of PCC pavements caused by the pumping actions under heavy wheel impacts. In 1974 diamond grinding of PCC pavements cost approximately \$1.75/yd$^2$, or about \$12,000/mile for one 12-ft-wide lane. Grinding is usually done in conjunction with other treatments; thus its cost is usually only part of the total outlay for restoring or improving a pavement.

Figure 8.6 shows a 12-year-old interstate highway pavement that has had its truck lane diamond ground. This photograph is at the high side of a superelevated curve in a location where severe faulting had oc-

**FIG. 8.6** 12-year-old interstate highway has had its truck lane (on high side of super-elevated curve) diamond ground to reduce severe faulting; poor drainage; no groundwater; West Coast area; April 1972.

curred, showing that the higher elevation, of itself, is not beneficial unless there is an adequate drainage layer under a pavement.

### Digging Out and Replacing Failed Portions

When heavy-volume segments of expressways or freeways show distress under traffic, it often becomes necessary to dig out and replace intermittent sections that have failed to the point where they are hazardous to the motoring public. Already (in 1974) rather large amounts of reconstruction and/or replacement of damaged interstate pavements has been required (see reference to GAO report in Sec. 8.1). Studies by the Highway Research Board indicate that in the next decade ". . . an important percentage of the pavement structures of the most heavily traveled freeways in urban areas will reach a condition where structural rehabilitation will be required to keep them operable" (see Project Statement for Research Project 14-4, (Highway Research Board, 1973). In the research problems statement for that NCHRP research project it was brought out that

> The most commonly used repair methods of patching, removal and reconstruction, and the placing of thick overlays over the entire roadway are time consuming, disruptive of traffic flow due to blockage in the construction area,

and generators of construction traffic for further interference. Patching rarely offers more than temporary relief. Removal and replacement has the additional disadvantage of wasting large quantities of existing pavement materials. Thick bituminous concrete and portland cement concrete overlays require wasteful coverage of the entire roadway, although only a portion of the width may be structurally deficient. Vertical clearance requirements offer a further restraint in the use of thick overlays.

Also that statement says

The duration and extent of interference to continuing use of the freeway facility during construction with current methods often create an impasse. Increased hazard to workmen and motorists makes timely, durable rehabilitation virtually impossible without closing or unacceptably restricting the freeway. . . .

Because of the very difficult and restrictive conditions under which broken sections of busy highways and airfields have to be dug out and replaced, the costs are quite high. Reconstruction of unreinforced PCC pavements in the west with high-early strength concrete backfill has been reported to cost around \$30.00 to \$50.00/yd$^2$ in 1974. The reconstruction of reinforced portland cement concrete in the central states has cost up to \$80.00/yd$^2$ in 1974. Figure 8.7, taken in a western

**FIG. 8.7** Fast-setting PCC replacement of failed section of 12-year-old interstate highway; West Coast area; December 1970.

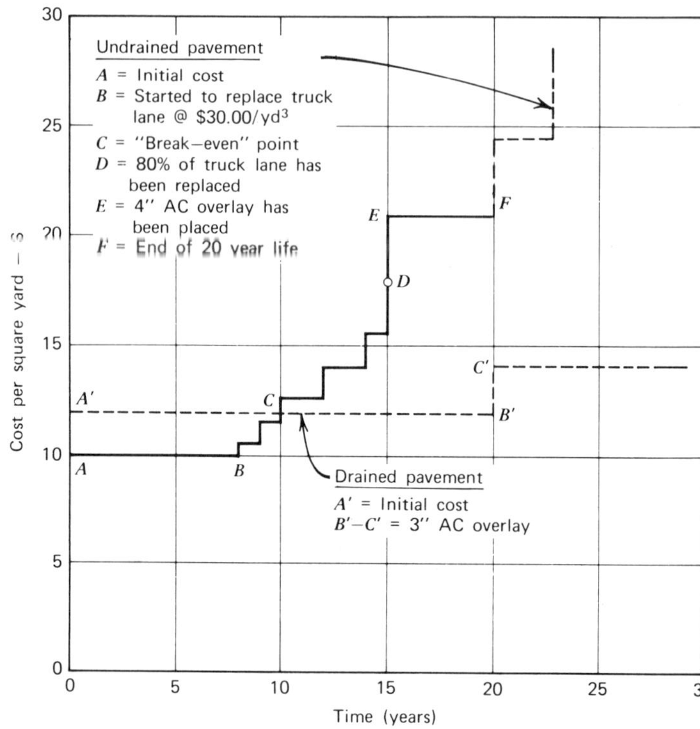


**FIG. 8.8** Relative costs of undrained and drained PCC pavements; an illustrative example.

state, shows a section of the truck lane on a 12-year old interstate highway that has been dug out and reconstructed with fast-setting PCC.

Figure 8.8 illustrates the way extensive digging out and replacement of unreinforced PCC pavements on freeways can lead to total costs far exceeding those of building well-drained pavements initially. Both the standard undrained and the drained design are assumed to have a 9-in.-thick PCC surfacing. The standard section has a 6-in. CTB and a 6-in. aggregate subbase for a total initial construction cost (exclusive of grading, embankment, shoulders, and other common features) of $10.00/yd^2$. The drained section has a 5-in-thick open-graded AC base drainage layer under the PCC, a 6-in. aggregate subbase and collector pipes and outlet pipes, for a total initial construction cost of $11.40/yd^2$ (see Example 1, Sec. 8.2).

Figure 8.8 gives the total capital outlays over an assumed 20-year life of the undrained pavement and likewise for the drained pavement having a greater useful life. The cost line for the undrained design follows *A-B-C-D-E-F*. At 8 years (point *B*), 5 percent of the truck lane (of the three lanes going in one direction) is dug out and replaced at a cost of \$30.00/yd$^2$. The cost per square yard of a three-lane roadway is (0.05) (\$30.00) (⅓) = \$0.50. Intermittently additional portions of the truck lane are dug out, and at 15 years the total amount is 80 percent of the truck lane (point *D*). At this time the cost per square yard of a three-lane roadway is (0.80) (\$30.00) (⅓) = \$8.00/yd$^2$ for restorations, for a total capital outlay of \$10.00 + \$8.00 = \$18.00/yd$^2$. At this point it is assumed a 4-in.-thick AC overlay is needed on the entire three lanes, for an added cost of (4) (\$0.75) = \$3.00/yd$^2$. At this point the total outlay has been \$18.00 + \$3.00 = \$21.00/yd$^2$. In Fig. 8.8 the cost line for the drained design follows *A′-B′-C′*. Since this pavement has a full-width macadam type of base drainage layer with pipe outlets and collectors, it is considered improbable that any abnormal maintenance will be required, but for general rehabilitation a 3-in.-thick AC overlay will be placed at 20 years for a cost of (3) (\$0.75) = \$2.25/yd$^2$, to raise the total cost to \$11.40 + \$2.25 = \$13.65/yd$^2$, after which it is felt that a good many additional years of service will still be provided by this pavement.

At the end of 20 years the drained pavement in this example will have cost (\$13.65/\$21.00) = 0.65 or 65 percent of the undrained pavement. Also, it is possible that within a few more years the undrained pavement will have to be dug out and completely rebuilt, which would make the potential savings from drainage even more spectacular.

Figure 8.8 is considered a graphic example of the potential benefits of drained pavements. Similar charts can be readily developed for any pavements for which cost data are available, or they can be reasonably estimated. If highly efficient drainage can reduce maintenance and replacement costs even a moderate amount, the small added cost of drainage will soon be more than repaid. Actual experience records of many heavy-duty pavements are proving this to be the case.

## 8.6 CONCLUSIONS

In this chapter, various aspects of the economics of undrained and drained pavements have been presented. It has been shown that well-drained pavements which provide longer, more trouble-free service than their poorly drained counterparts, are less costly in the long run. Well-drained designs require smaller quantities of vital construction materials over their effective lifetime; hence good drainage is a way to conserve dwindling supplies of cement, asphalt, and mineral aggregates.

The large costs for repairing and digging out and replacing portions of heavily traveled pavements, together with the many "hidden" costs of down time of runways and taxiways of airfields and airports and lost time due to delays of highway users while repairs are being made, all add to the true cost of building undrained pavements.

When all factors are considered, well-drained pavements will almost always be considerably more cost effective than undrained pavements.

## References

Cedergren, H. R. (1967), "Grain Size vs. Dollars in Drainage Aggregates," *Civil Engineering,* ASCE, November 1967, pp. 62–63.

Cedergren, H. R. and W. R. Lovering (1968), "The Economics and Practicability of Layered Drains for Roadbeds," Highway Research *Record* No. 215, Highway Research Board; pp. 2–5.

Cedergren, H. R. (1974), "Drainage and Dewatering," Chapter 6 in *Foundation Engineering Handbook,* scheduled for publication in 1974 by Van Nostrand Reinhold Company, New York (edited by Drs. H. Y. Fang and Hans F. Winterkorn).

Civil Engineering-ASCE (1970), "Interstate Pavements Failing Early," August 1970, p. 83.

Highway Research Board (1973), Project Statement for National Cooperative Highway Research Program Project 14-4, "Reconditioning Existing Heavy Duty Freeways in Urban Areas."

Grace, J. L., Jr., C. C. Calhoun, Jr., and D. N. Brown (1973), "Drainage and Erosion Control Facilities, Field Performance Investigation," Misc. Paper H-73-6, Waterways Experiment Station, Vicksburg, Mississippi, June 1973, p. xi.

*Rural and Urban Roads* (1972), "Better Subdrainage Can Cut Highway Maintenance," staff article, July, 1972, pp. 76–7.

CHAPTER NINE

# Performance of Open-Graded Materials

## 9.1 INTRODUCTION

Good drainage will help by increasing the life of pavements, greatly reducing the need for overlays and repairs and permitting many pavements to be constructed with less thickness of the paving layer than has been used in the past.

Since a layer of highly permeable, open-graded aggregate is the basic component of a subsurface drainage system for a pavement structural section, the performance record of such materials is of great interest. When an open-graded base drainage layer is substituted for part or all of a normally used slow-draining base, it has been considered that the substitution usually can be made on an "inch-for-inch" basis and that the drained design will be no thicker than the replaced undrained design. There is evidence, however, that drained designs may in many cases require from 10 to 30 percent less total thickness than structurally equivalent undrained designs. Each case should be evaluated on the basis of the structural capabilities of the materials making up the total cross section—both for normal undrained designs and for well-drained designs.

Open-graded aggregates, both for bases and for wearing courses, have been reported as providing excellent performance. When good-quality aggregates with a high percentage of fractured faces have been used in wearing courses, the performance has been superior to that of many of the normally used *dense-graded* materials. Some investigators say that elimination of the fines in mixes allows poorer quality aggregates to be used than is possible with the dense-graded mixes, thereby easing the demands on high-quality aggregates. Some engineers feel that the open-graded mixes can often be substituted on less that an "inch-for-inch" basis. Since 1 ton of open-graded material makes about 20 percent more pavement than 1 ton of dense-graded mix, the overall saving in vital paving materials might well be in the order of 25 to 50 percent when open-graded designs are used.

Carl W. Garbe of Dames and Moore, San Francisco consulting engineers, has recommended the use of open-graded crushed rock base courses as a way to save on materials required for several parking lots in the San Francisco Bay area. A parking lot for the Redwood City (California) corporation yard was designed with a 2½-in.-thick AC wearing course on a 6-in.-thick base course of 2 to ½ in. dia. crushed rock over a compacted subgrade with an $R$ value of 5.0. After 4 years of heavy-duty service this parking lot is in excellent condition. Its design permitted substantial savings in construction materials, because the conventional undrained design would have needed a total thickness in the order of 15 or 16 in. A key feature of this design, according to Mr. Garbe, was a close tolerance on the grading of the subgrade to prevent depressions that could trap water and positive slopes and valleys that feed the water to external drainage ditches (Garbe, 1974).

Although the ultimate benefit of the good drainage of these pavements will not be fully known until after many more years of service, a short-time experience with another pavement does give some immediate evidence of the value of positive drainage. This experience was described by Mr. Garbe as follows:

> In one instance that may be pertinent, however, the drainage of an open-graded base course under a small pavement area was blocked at its outer edge, and the AC pavement (2″ AC over 6″ crushed rock base) failed within a short time after construction. The repair was made with a concrete pavement over the same base (restored), and the subsequent successful service life is attributed (by some) to the rigidity of the concrete pavement. Measures that were implemented during the repair assured proper drainage of the base thereafter, and it is my view that the integrity of the new concrete pavement is the result of the drainage rather than the rigidity.

Well-drained dams and other hydraulic structures are safer and need considerably smaller amounts of cement and other construction materials than poorly drained structures. A well-drained concrete gravity dam, for example, uses only about 70 percent as much cement and aggregate as an equivalent undrained dam. When open-graded bases with positive collectors and outlets are provided for pavements, as proposed in this book, and rapid, effective drainage is assured, pavement life can be greatly extended, although minimal quantities of vital construction materials are required, thereby helping to conserve on natural resources such as cement, asphalt, and mineral aggregates. Well-drained, well-designed AC or PCC pavements should carry their design loads almost indefinitely with minimal maintenance. *It is considered likely that the use of open-graded base drainage layers can increase the efficiency of the most vital pavement materials by at least 25 to 50 percent, thereby allowing 33 to 100 percent more lane miles of pavements to be constructed with the same tonages of these materials as would be required using conventional designs.*

## 9.2 HISTORICAL RECORD OF OPEN-GRADED BASES

For centuries road builders have known of the good structural properties of well-interlocked layers of broken, durable stone. The ancient Romans built roads with thick layers of heavy stone, which probably depended more on massiveness than on interlocking. Since the 1800s, smaller-sized broken stone has been extensively used in road building. Thomas Telford, Pierre M. Tresaguet, and John L. McAdam are some of the leading road designers and builders of the past who developed improved road construction methods which made use of durable, broken stone. The macadam road was constructed by spreading a layer of crushed stone on a cushion of screenings or sand that had been placed on the subgrade to serve as an inverted choker course to keep soil from working up into the stone. After the crushed aggregate was placed and compacted thoroughly, a layer of smaller-sized material was added and rolled or vibrated until the upper stones were thoroughly keyed together. Various techniques were employed in the construction of pavements with macadam bases. To a large extent, stability was achieved through point to point contact of the crushed rocks (Bruce, 1934).

When the coarse, open-graded, macadam bases are properly drained, this kind of design can provide durable, strong roads, as evidenced by the success of this type of construction. But, if built in a trench with no provision for drainage, the open-graded rock simply allows water to flow to low areas where it collects in the structural section, allowing deterioration to set in. In addition, if the open-graded bases are placed directly on silty, sandy, or erodible clay subgrades without the choker layer, fines may in time work into the stone, rendering it weak and impermeable.

Richardson and Liddle (1973) list many outstanding benefits from the use of open-graded aggregates for forest service roads, which must support heavy logging trucks in wet climates in the Pacific Northwest (see Sec. 9.3). They point out that if fines are kept out of aggregates for emulsified asphalt pavements, there is less breakdown of the aggregates under heavy wheel loads, and that lower quality aggregates can be used than if fines are added and dense-graded mixes are used.

Durable, open-graded aggregate with a high percentage of fractured faces is one of the best materials available, for both wearing courses and open-graded base drainage layers. Several "case histories" of highway and airfield pavements in which open-graded aggregate was used are described in Sec. 9.5.

When open-graded base drainage layers are placed directly under the primary pavement course, the open-graded aggregate may be used without stabilizing additives if it is angular, durable material having sufficient structural strength to carry the design traffic. For ease of construction when rounded aggregate is used and in some cases for added cohesive strength, it may sometimes be advantageous to stabilize the layer with an asphaltic binder. Coyne (1972) points out that the open-graded mix proposed for a base drainage course differs from the original type of macadam construction, since the macadam road was generally built by placing a "... series of crushed stone layers, diminishing in size, each penetrated with asphalt cement. Open-graded mixes are combined by mixing rather than by penetration methods. Mixing assures a more uniform distribution of asphalt and minimizes the segregation problems inherent in placing open-graded rock."

One of the oldest known asphalt-treated, open-graded bases is a project which was constructed in Illinois in 1923 under the supervision of the late Gene Abson. The base was examined in 1963 by Mr. Abson, and was found to be in excellent condition (*Roads and Streets,* 1963) after 40 years of usage (see Case 4 in Sec. 9.5).

## 9.3 PERFORMANCE OF OPEN-GRADED PAVEMENT COURSES

Recent experience with logging roads and county roads in several states has demonstrated that open-graded aggregates stabilized with asphalt emulsions can carry surprisingly heavy loads without distress. Coyne (1972), who advocates the mixing of durable aggregates with cationic emulsions, "... to provide lasting stability. ..." points out that open-graded asphalt-treated mixes have a number of distinct advantages in road construction. The primary advantages he gives are the following:

1. Resistance to shoving and rutting under traffic.
2. Resistance to cracking under repeated load applications or thermal changes.
3. Reduced oxidation.
4. Adherence of the mix to the existing substrata.

Mr. Coyne says that, "Open-graded mixes are both permeable and erosion resistant and are ideally suited for layered drains for roadways. ..." He also states, "They are economical and durable. They offer the public both ecological and safety advantages. Their free drainage characteristics minimize the accumulation of water, thus reducing hydroplaning and debonding with the mix. . . ." (when used as a top course). Also, "The mixes are tough and abrasion resistant, yet flexible. Experience shows that open-graded mixes can resist reflection cracking better than dense graded mixes." And, "The principle design consideration is to provide sufficient binder to give the mix the desired flexibility and abrasion resistance."

Starting in 1966, several U.S. National Forest Service offices and counties in the Pacific Northwest have been using open-graded emulsified asphalt mixes to construct roads in remote forest areas. *With no wearing course except a thin chip seal*, many of these roads have been carrying large numbers of logging trucks weighing 100,000 to 200,000 pounds with remarkable success. The use of cold emulsions was initiated because of low construction cost, reduced air pollution, and as a means of conserving vital natural resources. Richardson and Liddle (1973) cite one project in which a conventional hot mix would have cost between \$7.00 and \$7.50/ton in place, but the open-graded emulsion mix cost only \$4.23/ton in place. They say that a simple pug mill suffices for the mixing, that there is no need for a dryer, and that the pavement can be put to traffic immediately. To achieve best results these authors recommend that the amount of material finer than the no. 10 sieve be held to a minimum (the less the better). They point out that large amounts of fine material cause an extremely fast break of the emulsion, and poor coating of the aggregate and that they require larger quantities of emulsion. The use of a minimum of fines permits thick asphalt films that give the mixes flexibility and long-lasting resistance to deterioration from oxidation. These authors say that *leaving the fines out conserves on supplies of mineral aggregates*, as ". . . more pavement can be constructed with each ton of mix."

Another benefit of the open-graded mix, according to Richardson and Liddle, is that, "Rutting and shoving which might be expected does not occur with the reduced amount of fines." Other advantages cited are (1) asphalt stripping caused by water or water vapor in the pavement does not occur because the open-graded condition allows free drainage, and (2) water cannot be pumped up from the subgrade, because without fines there can be no capillary action.

With regard to open-graded mixes containing very few fines or none at all, Richardson and Liddle also say,

> The second most important factor to the success of the mixture [the first is the low percentage of fines] is the inherent stability of the aggregate. This quality is dependent upon particle interlock as well as hardness and durability.

Crushed aggregates are therefore recommended for this reason. Some experiences with uncrushed river gravels have been less than encouraging.

These authors recommend a minimum of 60 to 75 percent of fractured faces.

These open gradations produce pavement wearing courses with 20 to 35 percent of air voids, which is a vital ingredient of open-graded emulsified asphalt pavements. Without fines, better stability is achieved than with denser pavement mixes, and rutting and shoving are eliminated.

Regarding the optimum amount of asphalt, Richardson and Liddle say, "... all the emulsified asphalt the aggregate will hold." They caution that the success of these mixes depends on suitable weather at the time of the construction. Mixing and paving can successfully be accomplished any time the air temperature is above 40 degrees, but "... rain is one weather condition which will produce disastrous results. Extreme caution is urged with open-graded emulsified asphalt paving when the threat of rain is imminent."

These writers go on to state,

> Although the open-graded emulsified asphalt paving concept defies many of the established theories of pavement design, its performance has been encouraging. Open-graded emulsified asphalt pavement has enough initial strength to support immediately vehicle loads without permanent damage to the pavement. The reduced pollution and the conservation of aggregates, coupled with the improvements in safety from the rough textured surface (and reduction in hydroplaning), are attractive benefits.

Altogether over 700 lane miles of these pavements had been constructed by 1973, often at savings in original costs of 25 percent or more.

These open-graded pavements evidently were looked upon primarily as highly efficient, economical wearing surfaces for the support of the unusually heavy wheel loads of logging trucks using many of these roads, with a side benefit of rapid drainage. In some localized areas where water has been trapped in these pavements, stability problems have developed.If these open-graded courses were to be provided with collector pipes and outlet pipes at every location in which water might otherwise become trapped, their benefits as drains could be more fully realized.

## 9.4 PHYSICAL PROPERTIES OF OPEN-GRADED MATERIALS

As noted in Sec. 9.3, when hard, durable aggregates are used in open-graded bases or pavement layers, high strength and durability are achieved primarily through the point-to-point contact between the particles. Leaving the fines out of mixtures not only increases permeability and drainability but usually increases stability and resistance to shoving and rutting while increasing flexibility and durability.

In 1972, the author cored with diamond drills through both PCC and AC pavements at a number of heavy-duty airfields in the United States. At one test location an AC pavement in the path of heavy planes (see Fig. 9.2) had been built on a base of crushed "trap rock" with no fines or binder. Although this pavement had been subjected to many years of heavy plane traffic, it was in excellent condition, with no evidence of cracking, settlement, rutting, or any other kind of damage. A base course of high quality crushed rock probably offers the finest type of support to heavy loads and heavy traffic volumes.

The excellent load-transmitting capabilities of coarse stone are often utilized in the construction of overlays over broken pavements to act as a cushion to prevent cracks from reflecting through the overlays. There is considerable evidence to support the conclusion that adequately designed AC or PCC pavements on durable macadam types of base drainage layers (with subbases as required) can be built with minimum quantities of vital materials such as cement and asphalt and can provide long-lasting performance with nearly zero maintenance.

In his study of the "optimum" properties of granular drainage layers for pavements, Winterkorn (1967) uses granulometric principles to evaluate the physical properties such as permeability and strength of various classes of granular mixtures. He concludes that "In drainage systems, our primary concern is the rate of flow of water." And, ". . . the size of the pores and hence the component particles should be as large as possible." He also says, "We must check, of course, the compatibility of these conclusions with the required mechanical strength properties of the system."

In his discussion of the findings of his study, Dr. Winterkorn states, "The obvious lesson for granular drainage structures is the use of single-sized coarse aggregate of maximum feasible dimensions. Of course, they will have to be protected externally by suitable filter layers." His overall conclusion is that

> Evaluation of available practical and theoretical information on the permeability and strength properties of granular systems makes it clear that these properties can be optimized by using the largest feasible aggregate of single size or very narrow size range. The aggregate should be of sufficient strength, toughness, abrasion resistance and durability to retain its integrity under service conditions, and its packing should be such that no further densification will occur.

Dr. Winterkorn's observations have been proved by the service records of highways, logging roads, parking lots, airfields, and the like that have been built with the use of these macadam types of aggregate mixtures. It is indeed fortunate that this kind of material is not only an optimum type from the strength and durability standpoint, but also from the drainage standpoint. Several examples of projects that have had outstanding performance with open-graded materials are described in the following paragraphs (Sec. 9.5).

## 9.5 CASE HISTORIES—PAVEMENTS ON OPEN-GRADED BASES

### Case 1—U.S. 99 (replaced by I-5) near Red Bluff, California

This 7.1-mile stretch of highway in northern California was constructed in 1947 with a $\frac{3}{4}$-in.-thick AC wearing course over an open-graded AC "binder course" $2\frac{1}{4}$ in. thick. After 10 years of heavy truck traffic, it was in such fine condition that it was presented as an outstanding job in the Asphalt Institute's *Quarterly* (October 1957) in an article entitled "Proven Asphalt Design." Its excellent condition is apparent in Fig. 9.1, a photograph taken after 10 years of traffic on this section of highway.

**FIG. 9.1** U.S. 99, near Red Bluff, California (replaced by I-5); 10-year-old pavement in excellent condition; has only $\frac{3}{4}$-in.-thick AC wearing course on $2\frac{1}{4}$-in.-thick open-graded "binder" course; 1957; *Asphalt Institute Quarterly*.

**TABLE 9.1** Aggregates Used in Red Bluff Section of U.S. 99 (now I-5)

| Course | Sieve size | Percent passing |
|---|---|---|
| Binder course (open-graded) | 1 in. | 100 |
| | $\frac{3}{4}$ in. | 90–100 |
| | $\frac{1}{2}$ in. | 35–65 |
| | No. 4 | 0–10 |
| | No. 200 | 0–2 |
| Surface course | $\frac{3}{8}$ in. | 100 |
| | No. 4 | 60–85 |
| | No. 8 | 40–65 |
| | No. 16 | 25–45 |
| | No. 200 | 2–8 |

The *Quarterly* article describes this road as follows:

In this rugged country state highway engineers in 1947 surfaced a 7.1 mile stretch of U.S. 99 with heavy-duty asphalt paving. Today this entire section of highway, near the town of Red Bluff, is providing superior, dependable service for the thousands of vehicles that travel it annually. According to maintenance superintendent E. R. Delano, in charge of the area, the pavement on the travel lanes has required practically no maintenance during the ten-year period.

It goes on to say

Engineers attribute this excellent service record to several significant design features. These include: (1) a pavement structure entirely adequate for the amount and weight of traffic carried during the ten-year period; (2) an open-graded hot-mix asphalt binder course that absorbs any possible faulting in the base; (3) side drainage of this open-graded mix that prevents the accumulation and ponding of water; and (4) a relatively thin hot-mix asphalt surface course supplying a tough, durable, waterproof roof over the underlying structure. This rugged, resilient carpet absorbs, without cracking, the deflections caused by heavy wheel loads.

This surfacing was constructed over a 15-in. subbase of clean, sandy gravel and a 6-in. cement-treated base of screened gravel. The asphalt concrete courses were constructed with aggregates of the gradations given in Table 9.1.

The binder course was hot mixed with 121–150 penetration paving grade asphalt and was spread at a rate of 210 lb/yd$^2$, to provide a compacted layer approximately 2.3 in. thick; over this layer the wearing course was laid at a rate of 60 lb/yd$^2$ to provide a compacted layer 0.7 in. thick, for a combined thickness of approximately 3 in. This wearing course was hot mixed with 5.5 percent 121–150 penetration paving-grade asphalt and was topped with a fine chip seal.

In summing up the fine performance of this section of "zero maintenance" pavement, the Asphalt Institute *Quarterly* article states, "The modern asphalt pavement of U.S. 99 has stood the test of a decade of heavy traffic with noteworthy success. . . ." The article concludes with the following summation: "Carefully designed with an eye toward future anticipated traffic, this highway should continue to give fine performance with no more than an occasional asphalt treatment or thin layer of economical hot-mix to strengthen it for increased traffic loads and maintain it indefinitely in top, smooth-riding condition."

The exceptional performance of the Red Bluff section of highway, as described in preceding paragraphs, is considered highly significant because it shows the way good drainage can assure maximum, trouble-free service with minimal amounts of paving materials.

### Case 2—Heavy-Duty Airfield Taxiway

For the most part, this taxiway and adjacent apron are heavy-duty PCC pavement (25-in. thick), but at one point the traffic line crosses a portion of AC pavement that contains a 6-in. layer of hard, durable, crushed rock with only a 6-in.-thick, dense-graded AC wearing course over it. The subbase is aggregate. This AC pavement has withstood

**FIG. 9.2** AC airfield taxiway pavement on 6-in.-thick "trap rock" base of high strength and high permeability; is in excellent condition (back left) after 18 years heavy plane traffic; November 1972. (Courtesy CERL.[1])

**FIG. 9.3** Sample of the "trap rock" taken from taxiway in Fig. 9.2 by core drilling through AC wearing course and into the base; core marks indicate high interlocking of base; 1973.

very heavy planes for more than 15 years with no signs of consolidation or structural weakness. Figure 9.2 shows this pavement (back, left) along the line of operation of heavy planes; it is nearly surrounded by the heavy-duty PCC pavement. When the crushed rock was cored into, it was so solidly interlocked that whole pieces of stone were completely cut through by the diamond drill without being dislodged. The coring marks on some of the pieces of stone are quite apparent in Fig. 9.3. This behavior is considered a vivid demonstration of the high internal strength of a compacted, durable, angular rock layer placed directly under a wearing course. Although the crushed rock layer was not provided with pipe outlets (as far as could be determined), it evidently bled water away from the heavy-load area to lower, outer places, leaving the primary pavement in a well-drained state. In a percolation test shown in progress in Fig. 9.4, it was not possible to build up any noticeable head in the coarse rock by pouring a pail of water into the hole as fast as possible. The coefficient of permeability of this layer was estimated to be in excess of 100,000 ft/day (35 cm/sec).

This pavement is considered to be an example of a relatively inexpensive yet extremely durable cross section, one that can sustain

**FIG. 9.4** Pouring water into test hole in pavement shown in Fig. 9.2; no head could be built up because of extremely high permeability of the base; November 1972. (Courtesy CERL.[1])

heavy truck or plane traffic with virtually no maintenance when adequately designed.

### Case 3—Kneeland Hill Road, Humboldt County, California

This short stretch of county road a few miles east of Eureka in the north coastal region of California has an AC wearing course over granular bases above a highly permeable macadam base drainage layer with a filter layer separating it from the subgrade. This special "experimental" pavement was reconstructed by the Humboldt County Public Works Department in 1967 after the original, poorly drained pavements had failed twice in just a few years from heavy traffic and excess water. The site is on Kneeland Hill in an area in which many springs and sidehill seepage are common occurrences, and a number of sections of the road have been damaged severely and dug out and repaired. This particular section of the two-lane road, on a steep 6.6 percent slope and a horizontal curve, was redesigned with the aid of W. R. Lovering of Sacramento, California, to utilize a highly permeable base drainage layer on a filter course to provide large water-removing capabilities and good filter protection against clogging from the subgrade soil.

As noted by Cedergren and Lovering (1968) the objectives of the experimental section were to find out if a highly permeable, open-graded base drainage layer would present any serious construction problems and to evaluate the water-removing capabilities of this drain. Subsequently this road was selected for one of the case study sites for the Federal Highway Administration's project for the development of "Guidelines for the Design of Subsurface Drainage Systems for Highway Structural Sections." When drilled into for this study (as noted elsewhere in this book) the structural layers were in a remarkably well-drained condition, and the pavement was in perfect condition (April 1971), after four wet winters.

When the "standard" types of local road building materials and drainage materials had been used in reconstructing this section of road prior to the 1967 reconstruction, excessive buildup of hydrostatic head in the structural section and failure occurred because the blanket of "standard" drainage material was incapable of removing the water that entered from springs and rainfall. According to the Highway Research Board article, "Pavement failure was the result, with water rising through cracks in the deteriorated pavement."

The 1967 reconstruction (after the two successive prior failures using normal materials) is considered by the author to be a noteworthy

**FIG. 9.5** Humboldt County's experimental pavement after testing for FHWA *Guidelines* project; its excellent condition is apparent from smooth, uniform appearance of the surface; April 1971. (Still in perfect condition in 1974).

**FIG. 9.6** Heavy flow of water out of drain pipe in Humboldt County's open-graded base drainage layer. Test made during construction (Courtesy of W. R. Lovering of Sacramento, California, who helped the county engineers design this section of road).

advancement in highway engineering. Although not large in size, it was a "bold" venture in the view of many of the local engineers who witnessed its construction. The excellent condition of this section of road at the time of the 1971 investigation for the FHWA project is apparent from the uniform condition of the pavement showing in the foreground in Fig. 9.5 (the dampness of part of this pavement in the background is from water used for cleaning the surface after the drilling for the FHWA study). Figure 9.6 shows a heavy flow of water coming out of the drain pipe at the lower end of this section in a test that was made during its reconstruction on the open-graded drainage layer in 1967.

### Case 4—Bush College Road, Illinois

*Roads and Streets* (1963) describes a short section of road between Illinois route 105 and Fairie Parkway in Decature Township, Illinois, that was constructed in 1923 by the late Gene Abson. After 40 years it was cored by Mr. Abson to find out why the road was still in such excellent condition. The pavement section consisted of a bituminous concrete base built in two layers, with an "open-graded" black base 5½ in. thick covered with a 1½-inch-thick sheet asphalt surface layer. After 40 years of service (5000 vehicles/day in 1963, including " . considerable heavy truck loads of gravel and soy beans") the inspection by Mr. Abson (cited in *Roads and Streets*, 1963) revealed ". . . that the roadway is in almost perfect condition, without depression, loss of contour or alignment and rides as smoothly as any newly constructed highway." The article goes on to explain, "Cores were taken from several locations. Visual examination of the cores showed an excellent state of preservation, without any indication of stripping of the asphalt or other sign of deterioration." Tests on the upper sheet asphalt topping indicated that it was in a remarkable state of preservation, with little oxidation; however the black base was found to have aged to an extremely low penetration and high softening point. "Yet it has preserved its function as a water-proofing binder, and cementing agent. The asphalt used was the same as that used in the surface so the high void content *and thin films* must be considered as one of the main reasons for this hardening."

### Case 5—Circular Test Track Experiments

In 1973 Dr. Ernest J. Barenberg supervised accelerated traffic tests at the University of Illinois for the Illinois Department of Transportation (Barenberg and Tayabji, 1974), in which valuable experience was gained with an open-graded base drainage layer over a silt subgrade. To obtain results in a comparatively short testing period, the designs were scaled down to produce definite overstressing of the structural sections. Six different designs were used. In all cases, a dense-graded AC wearing course with a thickness of 2.63 to 3.23 in. was placed over an open-graded bituminous aggregate mixture (designated OGBAM) using mineral aggregate between ¼ and 1 in. in size, stabilized with paving-grade asphalt. The thickness of the OGBAM was nominally 4 in., but varied from 3.68 to 4.11 in. as actually constructed. An A-6 silty subgrade was used under all test sections.

In test sections 1 and 6 a nominal 2-in. thick layer of lime-clay was placed between the silty subgrade and the layer of OGBAM; in sections 3 and 4, the OGBAM was placed directly on the A-6 subgrade; in section 2, an intervening layer of filter cloth was placed on the subgrade. Test section 5 had a layer of concrete sand between the subgrade and the OGBAM.

While applications with a 3300-lb. wheel load were being made, water was introduced at the high side of the section at scheduled intervals, and the quantity flowing through individual sections was periodically measured. Any significant pumping or other intrusion of soil into the OGBAM showed up as suspended soil in the effluent and in reduced quantities of flow; therefore any important breakdown of the system could be immediately detected.

Sections 2, 3, and 4 broke down after about 100,000 to 150,000 load applications, and sections 1 and 6 (lime stabilized) showed some evidence of rutting at the termination of the test, but no evidence of loss of drainage capacity or structural breakdown. Section 5 having the OGBAM on concrete sand gave the best performance, with little or no indication of severe distress to the time the test was ended (approximately 240,000 load applications). Because of the severity of the test, this is considered a significant demonstration of the excellent capabilities of open-graded drainage layers when placed on appropriate filter layers.

## 9.6 SUMMARY OF OPEN-GRADED BASE CASE HISTORIES

Information that has been presented in this chapter has demonstrated that all kinds of pavements, including both PCC and AC, can benefit from good drainage. The five examples in Section 9.5 are considered good illustrations of the fine field performance that can be expected of open-graded bases in pavement structural sections. The basic information about the first four cases, as summarized in Table 9.2, shows that a highly permeable, open-graded layer can be beneficial at nearly any level in a structural section. In the first example (Case 1), an open-graded AC layer *that is covered with less than an inch of dense-graded AC* withstood heavy truck traffic for 10 years and longer with no signs of any structural instability. (As noted in Sec. 9.3, open-graded emulsified asphalt mixes have withstood logging trucks weighing 100,000 to 200,000 lb in the Pacific Northwest, *with no wearing course at all except for a thin chip seal.* This is additional proof of the excellent structural properties of open-graded mixes.)

The second example (Case 2), a portion of a major airfield taxiway, shows that good quality crushed rock has superlative structural properties for heavy-load pavements—in fact, this type of design appears to offer outstanding advantages for nearly any pavements having to carry heavy wheel loads and large traffic volumes.

The third example (Case 3), is the Kneeland Hill road in Humboldt County, California, with a very permeable base drainage layer under

**TABLE 9.2** Summary of Open-Graded Base Case Histories

| Case | Site | Cross section | Comments |
|---|---|---|---|
| 1 | U.S. 99, near Red Bluff, California | 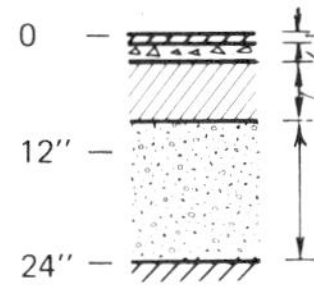 0.7″ dense AC<br>2.3″ open AC<br>6.0″ CTB<br>15″ aggregate base<br>Subgrade | *An outstanding example of excellent structural capabilities of open-graded AC*; only ¾ in. dense AC wearing course over open-graded binder course; in excellent condition after 10 years heavy traffic; (see Asphalt Institute *Quarterly*, October 1957) |
| 2 | Heavy-duty airfield taxiway. | 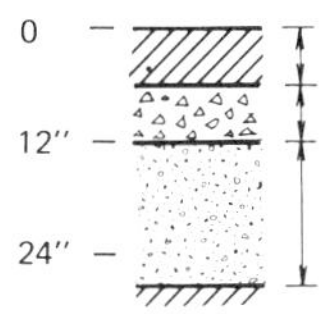 6″ dense AC<br>6″ crushed rock<br>15″ + aggregate base<br>Subgrade | *An outstanding example of superior structural capabilities of crushed rock bases*; relatively minimal section withstood over 15 years traffic of very heavy planes with no evident physical distress; probably can give excellent service indefinitely on busy highways, airfields, etc., with minor maintenance |
| 3 | Kneeland Hill Road, Humboldt County, California | 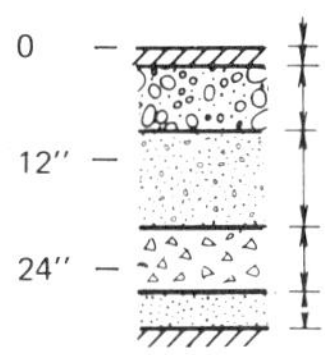 0.2′ dense AC<br>0.6′ aggregate base<br>0.9′ aggregate subbase<br>0.67′ open AC drain<br>0.33′ filter<br>Subgrade | *A very noteworthy installation* by county public works department (with aid of W.R. Lovering of Sacramento); open-graded AC drainage layer under primary section; has given remarkable protection from water; two prior repairs using "standard" filter layer failed quickly from excess water; heavy logging trucks use this road |
| 4 | Bush College Road, Illinois | 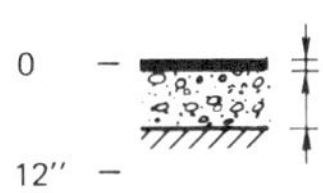 1.5″ sheet asphalt<br>5.5″ open AC<br>Subgrade | *A historic road*; built by Gene Abson in 1923; cored by him in 1963 after 40 years of moderate traffic; in excellent condition; Mr. Abson said that the open-graded AC "black base" had aged to a very low penetration, but "has preserved its function as a water-proofing binder and cementing agent" (see *Roads and Streets*, March 1963) |

base and subbase courses. When this pavement was drilled into in a study for the FHWA, all structural layers were found to be ". . . remarkably free of excess water," and were in fact only slightly damp, even though this pavement is in a high rainfall area with heavy sidehill seepage. Excess water and heavy loads had destroyed the previously constructed pavements within short periods of time, because the "standard" base courses and drainage materials provided negligible drainage.

The fourth example (Case 4) is Mr. Abson's road in Illinois, which was still giving good performance after 40 years of service. It is considered evidence of the long-lasting performance on an open-graded AC base.

The circular test track experiments of Barenberg (Case 5) demonstrate that a highly permeable open-graded base drainage layer can give excellent performance when a suitable subbase or filter layer is placed between the open-graded layer and a noncohesive silt subgrade.

## References

Asphalt Institute (1957), "Proven Asphalt Design," Asphalt Institute *Quarterly*, October 1957.

Barenberg, Ernest J. and Shiraz D. Tayabji (1974), "Evaluation of Typical Pavement Drainage Systems Using Open Graded Bituminous Aggregate Mixture Drainage Layers," Department of Civil Engineering, University of Illinois at Urbana-Champaign, for Illinois Department of Transportation, *Final Report*, May 1974.

Bruce, Arthur G. (1934), *Highway Design and Construction*, International Textbook Company, Scranton, Pennsylvania, pp. 289–295; 345–353.

Cedergren, H. R., and W. R. Lovering (1968), "The Economics and Practicability of Layered Drains for Roadbeds," Highway Research *Record* No. 215.

Coyne, L. D. (1972), "Design and Construction of Emulsified Asphalt Open-Graded Mixes and Overlays," Chevron Asphalt Co., 23rd Annual Road Builders Clinic, Moscow, Idaho, March 17, 1972.

Garbe, Carl W. (1974), private communication with author.

Richardson, Emory S. and William A. Liddle (1973), "Open-Graded Emulsified Asphalt Pavements," Region 10 Office of Federal Highway Projects, Implementation Division, Office of Development, Federal Highway Administration, Washington, D.C., June 1973.

*Roads and Streets* (1963), "What Happens to Black Base After 40 Years?" March 1963, pp. 130–131.

Winterkorn, Hans (1967), "Application of Granulometric Principles for Optimization of Strength and Permeability of Granular Drainage Structures," Highway Research *Record* No. 203, Highway Research Board.

CHAPTER TEN

# Hydroplaning Reduction with Porous Overlays and Pavements

Problems with hydroplaning have been reported on highways and airfields for many years. Likewise, rainy weather frequently causes serious hazards, delays, and so forth of exhibition racing at motor speedways, where major accidents have been at least partially attributed to water on pavements. In the past 5 or 6 years diamond grooving of both highways and airfields has been a fairly common treatment where hydroplaning tendencies have made pavements hazardous to the users. In 1974, grooving cost between \$1.25 and \$2.00/yd$^2$, which requires a substantial outlay if a large amount of pavement area is involved.

For many years a "popcorn" type of mix has been used on highways to provide a skid-proof wearing surface and to reduce hydroplaning problems caused by excess water films on pavement surfaces. Recently it has been employed on airfields.

Another type of highly permeable open-graded pavement (see Sec. 9.3), or the highly permeable open-graded base drainage layers described in this book, can also reduce the tendencies for hydroplaning by removing at least part of the free water that lands on pavement surfaces.

## 10.1 POPCORN FRICTION OVERLAYS

Even though the so-called popcorn AC overlays have been used on highway pavements since about 1950, it was not until fairly recently that they have been used on airfields. Jones (1973) tells about the use of an open-graded AC "friction overlay" that was placed on Runway 17-35 of the U.S. Naval Air Station at Dallas, Texas. Problems with hydroplaning had been occurring " . . . at an alarming rate in the past three or four years. . . . " on this runway. Pilot complaints of " . . . poor

**TABLE 10.1** Aggregates Used in Porous AC Overlay on Runway 17-35, U.S. Naval Air Station, Dallas, Texas

| U.S. sieve size or number | Percent passing |
|---|---|
| $\frac{1}{2}$ in. | 100 |
| $\frac{3}{8}$ in. | 90–100 |
| No. 4 | 30–55 |
| No. 10 | 0–22 |
| No. 40 | 0–12 |
| No. 200 | 0–5 |

From *"Civil Engineering-ASCE"*, March 1973, p. 46.

braking and loss of control on the wet runway created concern for aircraft safety. . . . " Flat cross slopes and poor surface geometrics were blamed for the slow drainage of surface water off this runway after a rainstorm. Also, the runway was shorter than usual, which increased braking problems.

According to Mr. Jones, comparisons were made between the merits of diamond grooving this runway or placing a coarse-textured overlay to improve surface drainage and reduce hydroplaning. Because the runway had been previously treated with a slurry seal, there was concern over the tearing of the surface by grooving and the potential creation of loose material on the surface if this method were used; thus the overlay was applied. This porous friction course was completed at a unit cost of $0.57/yd$^2$, or approximately $18/ton for the ⅝-in.-thick (compacted) overlay. When this overlay was made, Mr. Jones explains that a tack coat of less than 0.1 gal/yd$^2$ of RS-1 was used in advance of the lay-down machine, which is a conventional paving machine.

The aggregate for this overlay was crushed basalt, with a percentage of wear in the Los Angeles abrasion test of 15.4. The aggregate for the overlay conformed to the grading specification given in Table 10.1.

The actual material used was very low in fines and a very porous mix was obtained. The photographs in Fig. 10.1 show the appearance of this runway an hour after a rain *before* the overlay was placed (Fig. 10.1*a*) and an hour after a rain *after* the overlay was placed (Fig. 10.1*b*). According to Mr. Jones, complaints of loss of skid resistance immediately ceased after the overlay was applied.

**FIG. 10.1** Runway 17-35, U.S. Naval Air Station, Dallas, Texas, showing the effect of a popcorn friction overlay on appearance an hour after a rain (Courtesy of U.S. Navy and *Civil Engineering, ASCE*). (*a*) Appearance *before* placement of the overlay, (*b*) Appearance *after* placement of the overlay.

**TABLE 10.2** Rates of Rainfall that Can Be Removed by 0.05-ft-thick Popcorn Overlays*

| Coefficient of permeability (ft/day) | Quantity per foot ($ft^3$/day) | Rainfall rate (in/hour) |
|---|---|---|
| 1000 | 0.5 | 0.012 |
| 5000 | 2.5 | 0.06 |
| 10,000 | 5.0 | 0.12 |
| 20,000 | 10.0 | 0.24 |

* Applies to 1-ft strip of pavement with sloping distance of 40 feet; slope in direction of flow = 0.01.

Part of the effectiveness of these coarse-textured, porous overlays may be in a higher skid resistance of the mixes used, but part is no doubt attributable to elimination or reduction of the thickness of sheets of free water between the pavements and the tires of the vehicles. An indication of the capabilities of porous overlays to remove water during and after rains can be obtained with Darcy's law $Q = kiA$. Assume, for example, that the cross slope of a highway or runway pavement is 0.01 and that such an overlay has a thickness of 0.05 ft (0.6 in.). Then, for various coefficients of permeability, the potential rate of flow of water in the layer and the equivalent in./hour rainfall on a 40-ft strip of pavement would be as shown in Table 10.2. These rates would apply to highways with sloping widths of 40 ft (with zero longitudinal grade), half widths of 80-ft-wide taxiways (likewise with zero longitudinal grade), and so on. It can be seen in Table 10.2 that a 0.05-ft-thick overlay with a permeability of 10,000 ft/day could remove about 0.1 in./hour of rainfall. Since $Q$ is proportional to layer thickness ($A$) and $k$, a 0.1-ft-thick layer with $k = 20{,}000$ ft/day would be able to remove about 0.4 in./hour, and a 0.2-ft-thick layer with $k = 25{,}000$ ft/day could remove around 1.0 in./hour. Thus, by utilizing relatively thick, coarse overlays rather large rainfall rates could be discharged. Their suitability for wearing courses would of course need to be verified.

## 10.2 POROUS PAVEMENTS ON OPEN-GRADED DRAINAGE LAYERS

Another approach to the hydroplaning problem is the use of a porous pavement over a drainage layer. Normally it is desired that pavement surfaces be kept as well sealed as possible to restrict the amounts of water and foreign matter that can enter structural sections and drainage systems. But in areas or situations where hydroplaning can create serious safety hazards, these hazards could be reduced by placing a porous asphaltic type of wearing course over a highly permeable drainage layer. If the wearing course and the drainage layer have sufficiently high permeabilities, appreciable amounts of surface water infiltration can be accommodated by such a system when suitable collector pipes and outlets are provided. Figure 5.3 shows that highly permeable base drainage layers (6 in. thick) can remove surface infiltration rates up to 2.5 in./hour for 100-ft-wide pavements on a slope of 0.01 and 1.0 in./hour for 250-ft-wide pavements on the same slope (sloping in one direction).

For the pavement surface to accept relatively large rainfall rates to relieve hydroplaning tendencies, the wearing course must have sufficiently high porosity and permeability or be sufficiently jointed to allow these rates to enter into the underlying base drainage layer. Very dense, impermeable paving materials and seal-coated pavements could not be expected to accept very much water. The water-removing capability of open-graded drainage layers has been discussed in considerable detail in other chapters. The capabilities of a surface course to percolate water downward into an open-graded drainage layer can also be estimated with Darcy's law $Q = kiA$. For vertically downward flow into a porous nonfilled layer the hydraulic gradient is 1.0; hence $Q = k(1.0)A = kA$. Then, for each unit of pavement area $A$ the possible rate of inflow $= q = k(1.0)$, or

$$q = k \tag{10.1}$$

For a pavement with a wearing course having a permeability of 10 ft/day, for example, the potential inflow rate would be 10 ft$^3$/ft$^2$/day or 120 in./day, which is 5 in./hour. For a pavement permeability of 20 ft/day, the potential rainfall infiltration rate would be 10 in./hour. Research would be needed to determine the kinds of wearing courses, types of mixes, etc. that would have the necessary strength and stability, while maintaining high effective permeability over long periods of time.

Any wearing courses that are depended on for absorption of surface water into underlying drainage layers would need special care to maintain sufficiently high levels of permeability to accomplish the desired result. It would be important that maintenance personnel diligently clean dirt and other matter off the surface to minimize clogging of these layers. And no seal coats or slurry seals should be applied to these areas. Although asphaltic materials tend to knead to a tighter less permeable condition under traffic, this action is partially nullified by the formation

FIG. 10.2 A very wet pavement during a rain; mist and spray can be a safety hazard, in addition to any skidding hazard.

of cracks and open joints which tend to allow relatively large amounts of water to enter most pavements (see Fig. 3.2). Any water that enters through cracks and joints in otherwise tight pavements is helping to relieve the hydroplaning problem as it is reducing the thickness of water films on the surface. The provision of an open-graded drainage system under many of the types of wearing courses in general use on highways, airfields, and so forth would reduce hydroplaning tendencies in proportion to the amount of water that could enter the drainage system.

Noticeable layers of water can collect on pavements that do not have any means for its rapid removal, either by surface drainage or by absorption. Figure 10.2 is a photograph of a very wet pavement. Other very wet pavements are shown in other chapters. Airfield pavements with large amounts of standing water are shown in Fig. 2.1 and Fig. 4.3*d*. Several very wet highway pavements may be seen in Fig. 4.3*a*, *b*, and *c*, and in Fig. 5.11.

## 10.3 SUMMARY COMMENTS

Although the pavement drainage systems containing a layer of highly permeable, open-graded aggregate were developed primarily as a means of reducing damage to pavements, increasing the efficiency of the materials used, and extending pavement life, a potential side benefit of these systems could be a reduction in hydroplaning hazards. By placing a suitable porous AC wearing course over a drainage layer (with pipes), significant amounts of surface water could be absorbed by pavements and discharged by the drainage layers as long as the wearing course and the drainage layer retained high levels of permeability. However, if pavements are constructed and maintained in such a way as to minimize the entry of surface water into structural sections (the normal way recommended in this book), a layer of popcorn AC mix (or diamond grooving) may in many cases be beneficial in reducing hydroplaning.

To assure permanent reduction of hydroplaning for porous pavements on open-graded bases, the permeability of the wearing course must be maintained at an adequate level. Impervious overlays, seal coats, slurry seals, and the like must not be used on these pavements. If foreign material such as wind-blown silt or sand, washed-on mud, or other matter is permitted to accumulate in large amounts and enter the pavement or open-graded base, permeabilities can be greatly reduced; consequently, maintenance operations would need to be geared to keeping such materials out of the wearing courses of these pavements. In any area in which large amounts of tire rubber collect on pavements, such as near the ends of runways, the hydroplaning protection could not be preserved.

The rapid removal of free water from the surfaces of highways, airfields, and motor speedways has taken on increasing importance in recent years, not only to reduce hydroplaning tendencies but to reduce the inconvenience and hazard of mist and spray caused by vehicles moving rapidly over wet pavements. The need for controlling the mist-and-spray problem was emphasized in the proceedings of the Permanent International Association Road Congress (1967). When a highly permeable, open-graded base drainage layer is located under a primary wearing course (with collector pipes and outlet pipes), this opens the way to the better utilization of paving materials and faster removal of excess water, both to reduce spray and to improve skid resistance during and immediately after heavy rainstorms.

## References

Jones, Michael P. (1973), "Friction Overlay Improves Runway Skid Resistance," *Civil Engineering,* ASCE, March 1973, pp. 45–48.

Permanent International Association Road Congress (1967), "Drainage—Great Britain," *Proceedings XIII,* Tokyo, No. 1–6, pp. 16–20.

# List of Tables

# List of Design-Oriented Charts and Drawings

# Author Index

# Subject Index